Stress,
Work Design,
and Productivity

WILEY SERIES ON
STUDIES IN OCCUPATIONAL STRESS

Series Editors

Professor Cary L. Cooper
Department of Management Sciences,
University of Manchester Institute
of Science and Technology

Professor S. V. Kasl
Department of Epidemiology,
School of Medicine,
Yale University

Stress, Work Design, and Productivity

WITHDRAWN

Edited by

E. N. Corlett

University of Nottingham

and

J. Richardson

University of Paris

JOHN WILEY & SONS

Chichester · New York · Brisbane · Toronto

British Library Cataloguing in Publication Data:

Stress, work design, and productivity.—
    (Wiley series on studies in occupational stress)
    1. Job stress
    I. Corlett, E. N. II. Richardson, J.
    158.7   HF5548.8   81-13075

    ISBN 0 471 28044 5

Typeset by Computacomp (UK) Ltd, Fort William, Scotland
and printed in Great Britain by The Pitman Press, Bath, Avon

Contributors

Ulf Aberg
Professor of Industrial Ergonomics, Department of Industrial Ergonomics, Royal Institute of Technology, Stockholm, Sweden

D. Bosman
Department of Electrical Engineering, Twente University of Technology, Enschede, Netherlands

Cary L. Cooper
Professor of Management Education, Department of Management Sciences, University of Manchester Institute of Science and Technology, Manchester, UK

E. Nigel Corlett
Editor
Professor of Production Engineering, Department of Production Engineering and Production Management, University of Nottingham, Nottingham, UK

Tom Cox
Lecturer in Psychology. Director, Stress Research, Department of Psychology, University of Nottingham, Nottingham, UK

Kenneth D. Eason
Department of Human Sciences, University of Technology, Loughborough, UK

Francis Jankovsky
Laboratoire de Physiologie du Travail et d'Ergonomie, Conservatoire National des Arts et Metiers, Paris, France

John W. H. Kalsbeek
Department of Systems, University of Twente, Enschede, Netherlands

Osmo Karhu
Ovako Oy, Helsinki, Finland

Ilkka Kuorinka
Institute of Occupational Health, Helsinki, Finland

J. M. Leduc
Federation Generale de la Metallurgie, CFDT, Paris, France

Heinz Leymann
Researcher, Work Psychology Unit, National Board of Occupational Safety and Health, Stockholm, Sweden

Colin Mackay   *EMAS, Health & Safety Executive, London, UK*
Ilija Manenica   *Department of Psychology, Faculty of Science and
                            Art, Zadar, Yugoslavia*
Judi Marshall   *Lecturer in Organizational Behaviour, School of
                            Management, University of Bath, Bath, UK*
Pekka Oja   *Institute of Occupational Health, Helsinki, Finland*
Wolfram Scheibe   *Institut für Arbeitswissenschaft, Technische
                            Hochschule, Darmstadt, Germany*
Reginald G. Sell   *Work Research Unit, Department of Employment,
                            London, UK*
William B. Spry   *Consultant Psychiatrist, Bryn-y-Neuadd Hospital,
                            Llanfairfechan, Wales*
Helmut Strasser   *Private-Dozent and head of Working group, Institut
                            für Arbeitsphysiologie, Technical University Munich,
                            Munich, Germany*
Robert J. Tinning   *Former Project Officer, Human Factors Research,
                            British Steel Corporation, 74 Radnor Road, Harrow,
                            UK*

Contents

Editorial Foreword to the Series

This book, *Stress, Work Design, and Productivity*, is the fourth book in the series of *Studies in Occupational Stress*. The main objective of this series of books is to bring together the leading international psychologists and occupational health researchers to report on their work on various aspects of occupational stress and health. The series will include a number of books on original research and theory in each of the areas described in the initial volume, such as Blue-Collar Stressors, The Interface Between the Work Environment and the Family, Individual Differences in Stress Reactions, The Person–Environment Fit Model, Behavioural Modification and Stress Reduction, Stress and the Socio-technical Environment, The Stressful Effects of Retirement and Unemployment, and many other topics of interest in understanding stress in the workplace.

We hope these books will appeal to a broad spectrum of readers—to academic researchers and postgraduate students in applied and occupational psychology and sociology, occupational medicine, management, personnel, etc., and to practitioners working in industry and the occupational medical field, mental health specialists, social workers, personnel officers, and others interested in the health of the individual worker.

CARY L. COOPER,
*University of Manchester Institute of
Science and Technology (UK)*
STANISLAV V. KASL,
Yale University

Preface

The design of a machine or a process usually emphasizes the importance of technical efficiency but rarely sees the influence on that efficiency of the people in the system. What people do arises, almost by accident, from what the machinery does not do. What is more, not only is the content of the person's job often decided by convention or in other implicit ways, but the workplace in which he does it may also owe more to the convenience of the machinery than the operator.

Increased pressures on designers arise today as a result of increased concern with energy use, pollution risks from industrial processes, and environmental hazards of many kinds. At an early stage in the design process these are taken into account, particularly when the design specification includes clear requirements in these areas. It is usual for the definition of these hazards to be accompanied by design guidelines for their reduction, drawn from scientific findings arising from experience and experiment.

For some thirty years there has been experience and experiment in the field of ergonomics, and findings from work physiology, applied psychology, and industrial sociology have been translated into norms for the design of human work. However, whilst the backgrounds of the physical scientist and the engineer are broadly similar and thus enable the findings from one to be utilized by the other, that of the human scientist is less so. His methods are less readily useable by the engineer, as are his criteria, so the introduction of the norms is inhibited by a lack of methods, as well as of data in an appropriate form.

To develop these requirements means encountering several problems. Major amongst them are the following:

1. During the conventional design process a clear insight into the actual content of jobs and the relationships between the requirements of workers can only be obtained at a late stage in design.

2. As mentioned above, the ergonomic norms may be difficult to translate into engineering terms.
3. As a consequence of the previous point, engineering designs which will meet ergonomic norms may need innovation, both technical *and* organizational.

Fortunately, systems are rarely entirely new. Most are predominantly the result of development, and it is possible to learn much from studying the weaknesses of earlier generations of systems. Even when this is done, however, the findings require presentation in a format useable by the designer. This is not easy when many engineers have scanty knowledge of human abilities and limitations. Even today it is probably the minority of engineering students which hears about ergonomics during training. This presents a severe limitation on the improvement of working conditions and leads to increased costs when social demands require the modification of existing equipment.

It is worth recognizing that technical progress is not an inevitable and uncotrolled process. It can be overlooked that the direction of technology is only one of several possible directions, chosen as a result of the economic conditions and constraints of the time. At most stages of technological development some freedom exists to choose from a range of possibilities and the choice depends on the balance of the criteria which have force with the engineer or manager at the time. A change in the balance of forces would lead to a change in direction of the technology and our present society is only one from a number of feasible and viable societies which might exist today.

To extend this point in relation to current technical developments, consider the enormous increase in freedom for designers given by microelectronics. They permit a wider choice of criteria than just the economic, the efficient and the reliable performance of the product, since they enable the introduction of many more functions than was formerly possible. However, systems are still developed with these three, admittedly very important criteria as overriding, with the resultant jobs decided implicitly and with no consideration of the work experience itself. The current state of the technology makes it quite feasible to define, and include in the specification, the job content and organizational requirements for satisfactory work. Software developed with such criteria in mind, the production of which is always a costly component, would not then limit the job quality for years to come but could aim directly at the creation of jobs which could be done and would be acceptable to modern workers.

The blind acceptance of only the classical criteria for industrial machinery leads us towards total automation, a solution which is not necessarily universally satisfactory. Problems of employment and maintaining the social organization in the so-called Third World require industrial solutions tailored to their social needs. This is equally true of the varied nations of Europe and America. It is the purpose of PROMSTRA, not to push for any particular solution but to present and discuss practical alternatives to designs based only on cost criteria. The objective is to assist industries to seek for themselves solutions to their production problems which will maintain their productive performance with the willing cooperation of members of a modern society.

J. W. H. KALSBEEK,
Founder President of PROMSTRA

Introduction

In the early years of the 1970s a small group of European academics came together to discuss their interests. They were all scientists, engaged in the study of industrial work and workers, and all had experience of working in industry as well as in the laboratory. Their concern was with the more widespread recognition of the importance and value of the results of research into industrial work which had been an increasing European field of investigation since the end of the Second World War. In particular, they were concerned with the more rapid adoption by industry of the results of this research.

The anxiety for its increased utilization sprang from several concerns. Firstly, efficient and internationally competitive industry is necessary for Western Europe; secondly, the increasing numbers of workers from developing areas coming to work in European industry indicated both a comparative decline in interest in industrial work by the native population and a potential source of social problems for the future; thirdly, growing demands for a higher quality of provision for health and safety in industry appeared to conflict with a need for even higher productivity in the view of some industrial managers; fourthly, but by no means least amongst the concerns, the increasing body of knowledge concerning effective and humane job design seemed to be having no visible effect on the generality of companies, whose jobs seemed to be locked into a pattern more relevant to the first quarter of this century than the last.

The methods selected by this group to put across the information being gathered concerning work design were twofold. They believed that the information would transfer best by face-to-face discussion with industrial people, be they managers, engineers, workers or union officials, during which particular problems would be explored. The material which follows has arisen during such presentation, when small groups of about thirty production managers and academics met for three or four days to discuss ideas and experiences in the development of new work practices. The informality, small numbers and relatively unstructured form of the seminars allowed the industrial participants to explore matters of interest to them rather than those questions put forward by the organizers. Thus problems of implementation could be explored in depth, theory taking its place as the support for and explanation of the practical events rather than the other way round, which is the more usual situation.

The second method adopted by the Promstra Group is the publication of this book. Its contents represent a selection from several seminars. The seminars each had a separate focus, such as assembly work, repetitive machine work, industrial safety, and office work. The main purpose underlying the book is to provide the interested industrial manager or engineer with evidence of the need for improved human work, the evidence that it can be done within the context of a profitable operation, indeed that the betterment of jobs is in itself profitable. It is also a function of the book that some ideas and methods for changing work are given so that there is a variety of starting points and directions available to the reader.

So much for the immediate parentage of this book. Its coverage is far from comprehensive for it touches the immense problem of considering the development of an industrial structure appropriate for the future of man at disparate points. The selection of these points has been by the arbitrary procedure of dealing with those problems which are of current concern. These points do not yet define the total surface of a new technological relationship for man and we must still rely on philosophical formulations of this relationship to guide us in attempts to create better working situations.

The starting points of these various formulations will influence their structure, yet many of them result in a clear recognition that the near future is likely to demand dramatic changes in society. Sometimes these will arise from shortages of physical needs, some suggest the pressures resulting from Thrid World deprivations or First World greed. Others recognize an inevitable trend in social aspirations and relationships in industrial countries.

A group which has made a major impression on thinking about the likely shape of the future are those responsible for the Club of Rome study (1974). They predicted the absence of certain essential minerals at given future times and then presented survival scenarios which might arise as a result, together with proposals to forestall their arrival. Another author (Schon, 1971) suggests a dramatic change in the nature of society, proposing that unstable social relationships will be the norm and society will adapt to instability. Most such discussions base their predictions on extrapolations from the relatively recent past. If the availability of a mineral is being predicted there is little doubt that its use in the near future, whilst it is still available substantially at its present price, will maintain its present growth pattern. Where human behaviour is under discussion there is less likelihood that sudden and permanent shifts in its manifestation will take place.

Biological adaptation is a slow process. The effects of adaptations forced by technology are usually seen in disorders of the organism. These may be controlled technically but result in new forms of disease and changes in the balance of fatal ailments, which remain until the causes are either removed or compensated by medical knowledge. In human terms, both physiologically and socially changes in ways of living are only slowly adapted to. Social organizations undergo transformations more rapidly than human physiological processes, but even these can be seen to be a matter of generations. It seems likely that each individual moves a step away from the position held by those of his previous generation rather than there being a continuous change in individuals over a lifetime. Whichever mechanism is present, the acceptance of new social structures and their effective operation can be a matter of many decades, a point not emphasized by politicians, who would prefer the electorate to believe that the millenium is only as far away as the next government.

If we accept that physical adaptation is a matter of centuries and social adaptation a question of decades, it follows that to take full advantage of the potential of technology we must use its flexibility to design it primarily for man, not for itself. If this is not done it destroys man, either directly or by destroying his social cohesion and thus the cooperative basis of his society. This conclusion, however, is not yet generally accepted. What is more usual is to push the technology as far as it will go, which means using the limited understanding which exists at the boundaries of knowledge, and fitting the resultant technology with palliatives to protect the user from many unknown, and some ignored, harmful aspects.

As an example we might consider the introduction of the computer, most recently in the form of the word processor, into the office. There are many groups, including ergonomists, working in software problems. These are frequently expressed as presenting to the operator information from the equipment in a form to minimize his errors and maximize his information intake in relation to the software. This software has been designed to utilize the technology to the full. But concern with the size of letters, screen contrast, glare or formatting are palliatives, necessary but secondary. The primary need is to design the word processor's facilities to permit the user to extend his abilities and exercise more control over his job. The former procedure is a more sophisticated form of Taylorism, the latter is enabling a mature adult to do a more efficient job of work.

The three sections of this book provide some theoretical considerations but try most of all to indicate how relatively unsophisticated methods can reveal some adverse effects of ordinary industrial or office work and, at the same time, show how improvements may be introduced. The book is offered as a step to assist workers at all jobs, whether they are managers, union officials, operators, designers or any other of the myriad range of industrial jobs, in the task of improving their working ability and their physical and mental health at the same time. In the words of Dr. R. Murray, when he was Medical Officer of the Trades Union Congress, 'Ergonomics is a new kind of game, in which both sides win'. There is no doubt that this is true. Work may never become preferable to play for the population as a whole, but it is certain that it does not need to be so depressingly and damagingly 'work'. A miniscule transfer of the resources used to develop technology to the problems of making it more workable by people would multiply the activities of those working in the field manyfold. We trust that this book will be an encouragement to those wishing to pursue the goal of better jobs, with its inevitable concomitant, a better world.

References

Meadows, D. H., Meadows, D. L., Randers, J., and Behrens III, W. W. (1972). *The Limits to Growth*. Pan Books Ltd., London.
Schon, D. A. (1971). *Beyond the Stable State*. Temple Smith, London.

PART I

Methods of Assessing Physiological Working Stress in Industry

Introduction

As discussed in the Preface, PROMSTRA is concerned with stress at work and the design of jobs to provide work loads which are appropriate for the people doing these jobs. Physical effort and its resulting feelings of fatigue or body discomfort are the most commonly experienced aspects of heavy work. It has been possible for many years to measure the 'cost' of heavy work to the worker, where 'cost' refers to the amount of energy required to do the work. The measurement of physiological 'cost' makes it possible to quantify levels and duration of work and to specify relaxation periods necessary for the worker to recover physically so that his health is not at risk.

It has been considerably more difficult to define what is physically taxing in tasks where the total energy cost is light but which require a repetitive set of gestures. This type of task is common in industry and is typified by the work on an assembly line. There is still a need for studies in this area which could provide methods of assessing light work and its effects on the operator's health.

Part I presents a variety of methods for assessing the effects of different physiological stresses found in industrial tasks. The chapters are not intended as an introduction to or survey of the whole area of physiological stress, but to demonstrate how certain methods developed from fundamental research can be adapted to assess physiological strain in industry. These methods provide an objective means of measuring the human organism's reaction to task demands. When used with subjective criteria of work load they should provide a reliable assessment of the physical load on the operator.

The production engineer or manager cannot be expected to carry out sophisticated physiological measures on the workers in his factory, and if the problems he is presented with are difficult it may be necessary to employ a physiologist or ergonomist to use these techniques. However, the chapters describe a variety of techniques, from sophisticated electrophysiological methods to a relatively simple observational method designed to be used by a non-specialist. The different levels of analysis of physiological parameters suggest a methodological approach which could be employed by the production engineer confronted by a problem of working conditions on the shop floor. The initial assessment of the situation could be made by using observational techniques applied by an employee of the firm who had been trained in their use. The first-level diagnosis would help towards formulating the problems in terms which could be presented to an ergonomist or work physiologist if it appeared necessary to continue studying further. The recent interest in working conditions in France has encouraged the use of unsophisticated methods of evaluation by non-specialist technicians, union representatives, and safety representatives to study the working environment in their own firms. This trend will probably continue as both sides of industry become more involved in providing a working environment adapted to the worker's needs. The methods referred to do not together provide the complete answer to evaluating working conditions and also suffer from a number of the drawbacks associated with checklist methods. However, they can be considered as a useful first approach to ergonomic problems which can be used by those persons directly involved in the work being studied.

The four chapters in Part I deal with the physical effort involved in working: they describe some measures of this effort and how the consequences of applying effort can influence the workers' production. The first two are concerned with effort resulting from physical movement or the exertion of forces to complete a task whilst the other two are particularly concerned with the effects of adopting different postures.

PROMSTRA's approach emphasizes the necessity of producing a practical approach to work design, and it is evident that any analytical techniques advocated by ergonomists will come up against long-established methods widely applied in industry. Manenica's chapter compares work study methods and direct and indirect physiological methods of evaluating physical workload by measuring the heart rate and oxygen consumption of forestry workers. He also attempts to formulate an index which requires minimum measures of the worker's physiological state. The study used industrial workers at their daily task, forestry workers in this case, who were trained both in terms of skill and of physical capacity and whose energy output could be considered higher than average. The index proposed gave more realistic measures of the effort required by the task and the rest pauses needed than conventional work study estimates. People at work do not take rest in the same way that work study calculations allow for it, i.e. as a proportion of each work cycle. It is more usual, and often more practical, to rest after a number of cycles, but this inevitably increases the proportion of rest needed if the work is heavy. Measuring the worker's response to work will take this into account and must be one of the reasons for discrepancies between the different methods.

Scheibe's chapter introduces a number of electrophysiological measures available for assessing physical work load. He summarizes the measurement techniques and discusses the limitations on their applicability under industrial working conditions. The technical limitations are becoming less and less important as instrumentation is developed to be more acceptable to the person being studied and the data analysis techniques allow large quantities of data to be handled effectively. Recent examples of long-term recording of heart rate of industrial workers have shown that it is feasible to use these techniques on the shop floor. However, the technical advances in recording have not necessarily been accompanied by similar advances in interpretative skills, and consequently electrophysiological records require assessment by a qualified person. This chapter describes how these techniques have been used to evaluate a particular workplace. It will be evident that the author's research team takes the perspective that the human organism functions in a complex manner and that it is necessary to monitor a number of variables. This is usually the case where 'light work' is being investigated.

The assessment of physical work load and how effort should be compensated by rest pauses poses considerable problems, not the least being the assumption that because a particular production level is being regularly achieved it must be considered satisfactory for the worker. Scheibe's chapter, by measuring exactly how the worker reacts biologically, shows that a slight increase in output rate during the working shift has definite effects on the heart rate. The explanation may well lie in the increased muscle activity, as it is known that an increase in output is not necessarily accompanied by a proportional increase in heart rate. The important point is that it should be recognized that working strains are not linearly related to work output and models of performance based on a linear relationship may seriously underestimate work load.

The effects of light repetitive work will probably not be recognized as a serious risk to health, as the individual parts of the task are not heavy and the work is perceived as comprising these individual and separate tasks. Only after a task analysis will all the individual components and their interactions be identified. A valid assessment still remains difficult, as it is not often recognized that apparently quite easy movements or postures are acceptable for short periods but can be fatiguing and even dangerous if repeated frequently over each shift for a whole working life. Redesigning to avoid this requires an understanding of the way the task requirements influence the posture and activities of the whole body over time and not just the immediate hand and eye movements related to the activity being performed.

Postures adopted by workers during their work also reflect the behavioural as well as the physical constraints of their tasks. The last two chapters in Part I discuss methods of evaluating postural stress at the workplace. Each procedure looks at movements and the resultant postures but is more concerned with the twists and loads on the body than the energy requirement to carry out the movement. A limitation in both methods is that they do not take into account the length of time for which each posture is maintained nor the sequence in which different postures are adopted. Both methods of evaluating postures involve a subjective judgement from the worker himself. Corlett asks the worker to comment on levels of discomfort over a working period, while Oja, Kuorinka and Karhu ask him to classify the strain of each posture he adopts. Corlett's chapter attempts to assess the relationships between posture and work performance in both dynamic and static industrial tasks. Oja *et al.* describe the development of a posture-recording technique designed as a tool for technicians and engineers in the Finnish steel industry. Epidemiological evidence has shown that the postures adopted by steel workers are largely responsible for the high incidence of musculo-skeletal problems in this industry. A simple assessment technique such as the one described here provides a means of identifying workplaces which demand difficult postures and will also allow new workplaces to be checked for postural stress before they are put into service.

The chapters in Part I are concerned with the question of working conditions and the assessment of physiological work load. The discussion on the difficulties associated with the use of the measures described points towards the need for research to provide viable methods for use in the field. These methods are not necessarily required to be as precise as laboratory measures but must be robust enough to be employed by non-qualified users in a variety of situations where laboratory controls are absent.

Stress, Work Design, and Productivity
Edited by E. N. Corlett and J. Richardson
© 1981 John Wiley & Sons Ltd

Chapter 1

Ergonomic Studies on Introduction of New Working Structures

Wolfram Scheibe
Institut für Arbeitswissenschaft, Darmstadt, Germany

This chapter is concerned with human reaction to stress, referred to here as 'strain' arising from working conditions. The assessment of the worker's reactions to stress in line work or any redesigned form of work requires appropriate variables to be measured throughout the work period, and some findings occurring during the project 'Introduction and Evolution of Improved Working Structures in Electrical Engineering in Germany' will be described.

Two main problem areas are associated with the scientific investigations for the introduction of new working structures:

1. The ergonomic consultation in the planning stage of the development of new working structures;
2. The comparison and evaluation of stress and strain in conventional and new working structures.

The term 'working structure' is conceived on the one hand as the static and dynamic relationships in a given working system in respect to its organizational, ergonomic and social aspects. On the other hand, the term 'work structuring' means the activity of the organization itself, the organization of the work situation and conditions, synchronizing the work content to the capacity and aspirations of the workers (Philips, 1968). Besides this process of mutual accommodation of human beings and work, the aim of our present ergonomic studies is to investigate to what extent the introduction of new working structures will bring about an improvement in comparison with conventional working structures.

To evaluate a working system's design by obtaining actual ergonomic results necessitates the performance of experimental investigations at the workplace. These 'ergonomic field studies' may well include physiological

7

measurements during the work processes. Although the analysis of our own investigations was not complete at the time of this report, the methodology and implementation of such an ergonomic field study will be described by means of the following example.

Figure 1.1 shows the assembly workplace for building ignition distributors at which the strain measurements were made on two women workers. From an ergonomic viewpoint the form of work characterized in this workplace can be defined as 'one-sided muscular work' or, following Murrell, 'active light work' (1971). The demarcation between static work and heavy dynamic muscular work can be made through the manner and frequency of muscular contraction. Therefore, factors which influence stress and strain can be found that provide the possibility of an evaluation of the working system's design conditions.

In this context it seems useful to mention briefly the definitions of the terms stress and strain. 'Stress' or 'load' is used for all factors of work which produce reactions in the person's receptor and effector systems. On the other

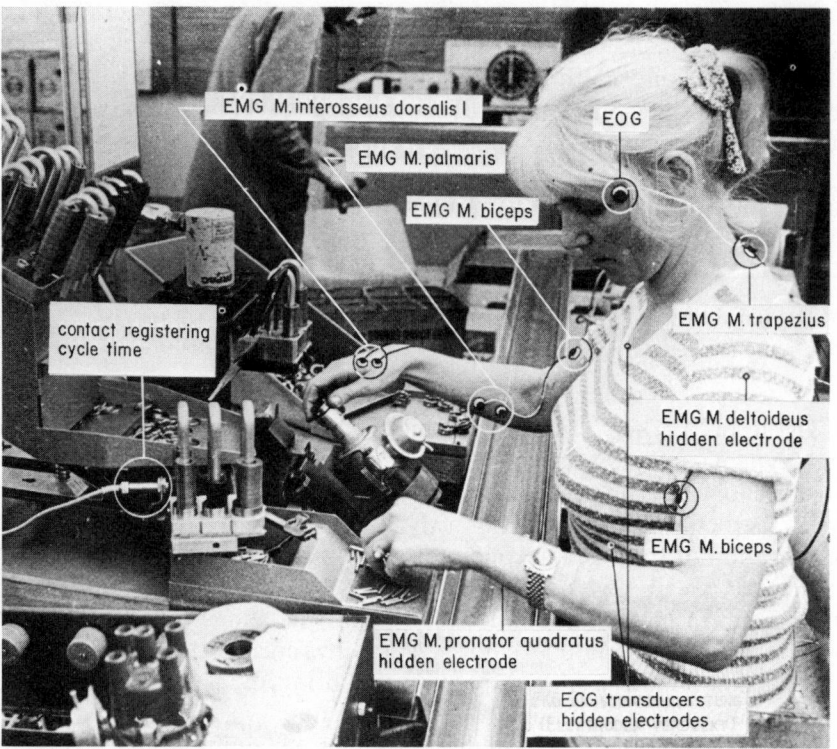

Figure 1.1   Assembly workplace and recorded variables of stress and strain

hand, we use the term 'strain' as the reaction to stress, so strain is the result of distinguishable stressors on the human being. Thus stressors or factors of stress are evaluated by analysing objective factors of the work or task, while strain is assessed by analysing suitable physiological or psychological variables (Rohmert et al., 1973).

The working cycle time and the running time during the whole shift were measured as factors of stress. The electrical impulses produced by special contacts on the equipment were recorded on magnetic tape; time periods could be calculated later from these and performance computed by appropriate digital data conversion equipment. The variables recorded to identify stress and strain are shown in Figure 1.1.

Heart rate, as an easily measured physiological indicator of strain, is a relevant measure for this one-sided muscular work (Rohmert, 1973). The intensity of its reaction is influenced by the working muscle, and in this case the reaction can also be influenced by emotional stress and by modification of body position.

Figure 1.1 shows the two surface electrodes for picking up the electrocardiogram from which the momentary heart rate was computed by time period measurement (Rohmert, 1973). In contrast to this, surface electromyography gives a physiological record directly coupled with the process of contraction. Electrical potentials from 10 muscles, which can be seen in Figure 1.1, were therefore picked up by electrodes at the body surface. They were then fed through suitable amplifiers and recorded on magnetic tape simultaneously with the electrical signals of cycle time. The so-called 'electrical muscular activity' was digitized by integrating the electromyogram in relation to the registered work cycles as time base. Electrical muscular activity is a valid indicator for local muscle strain.

A physiological variable used to describe the visual strain was the EOG (electrooculogram), which was picked up by surface electrodes on either side of the eye to record eye movements in the horizontal direction. These also were recorded on magnetic tape. By integrating over the work cycle periods the 'electrical eye activity' could be derived as a measure of the abundance and velocity of eye movements (Rohmert, 1973).

Figure 1.2 illustrates the layout of equipment at the workplace. Recording these variables of stress and strain simultaneously on magnetic tape resulted in a total sum of about fifty thousand measured values per shift. Further evaluation was therefore achieved by a combination of process and digital computer. By this means, data compression and synchronization were achieved, producing a matrix with strain and stress parameters. Their temporal variation could be described by statistical models of correlation and regression and parallel time plots of all variables.

The main results showed that the ergonomic design of the investigated work system should be improved if intolerable strain was to be avoided.

Figure 1.2   Layout of equipment at workplace

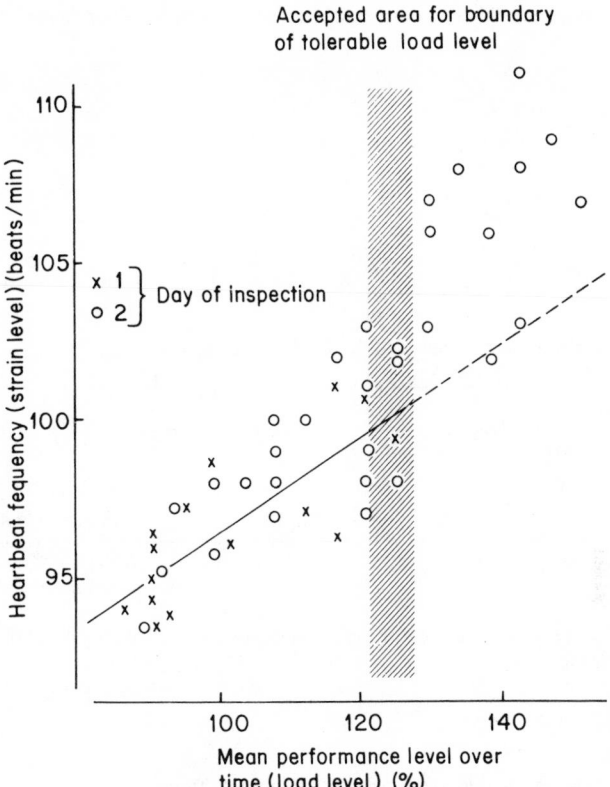

Figure 1.3   Relationship between stress and strain parameters expressed by performance and heart rate

The remaining diagrams, for instance, demonstrate the connection between recorded stress and strain parameters. Figure 1.3 shows the increase in heart rate with increasing performance. The dimension of heart rate is 'beats per minute', that of performance or output rate 'pieces assembled per minute', computed by inversion of the wage for piecework.

It is notable that the correlation between both variables is similar for both days. Above a certain degree of performance, the direct proportionality between level of stress, expressed by pieces assembled per minute, and level of strain, expressed by heart rate, changes. The subsequent nearly hyperbolic increase in heart rate indicates a deviation from the steady-state equivalent to a level of stress above the limit of tolerability.

Figure 1.4 illustrates the effect of practice from the temporal behaviour of electrical muscular activity and output rate, both shown in relation to the cumulative time of shift. In the upper diagram the beginning of the shift is

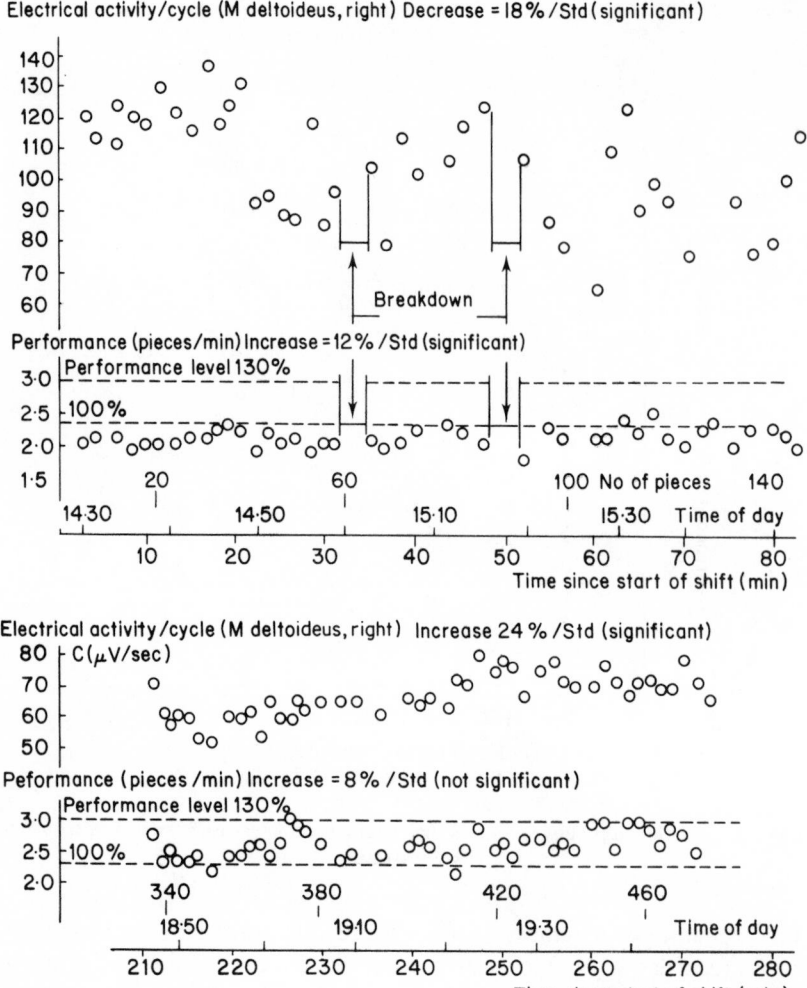

Figure 1.4 Absolute temporal variation of electrical muscular activity and performance (M. deltoideus)

recorded and shows decreasing electrical muscular activity in relation to a relatively low level of performance. At the end of the shift (lower diagram), for higher output rate a significant rise in the physiological strain indicator (electrical activity) is obvious.

The two levels of performance show a remarkable difference in deviation of the values. The muscle chosen is part of a group of muscles which fix the

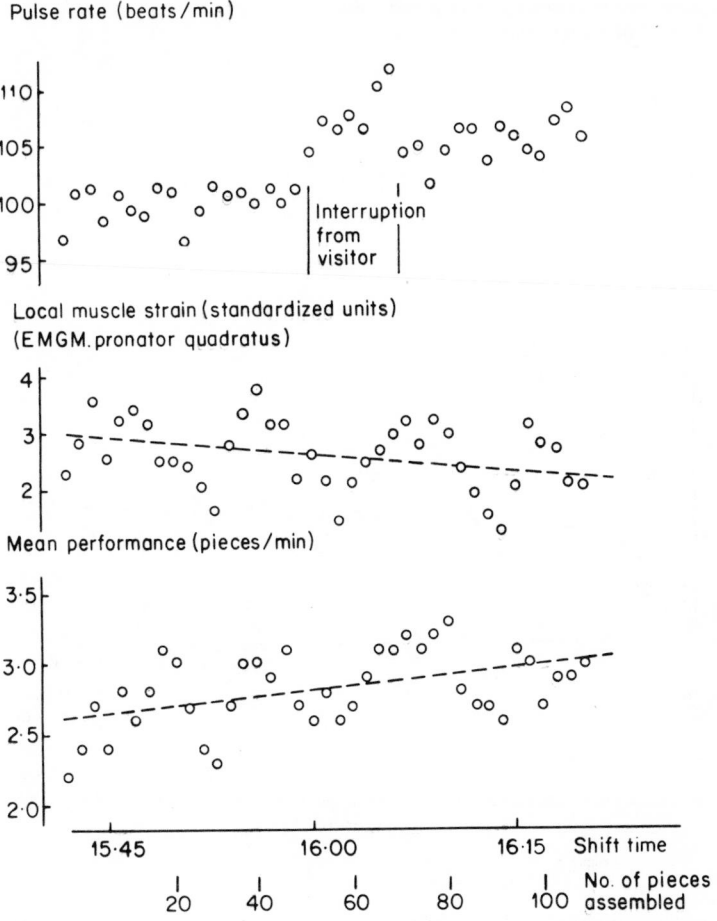

Figure 1.5   Temporal changes of strain parameters and performance at beginning of shift
(M. pronator quadratus)

position of the human body. Figures 1.5 and 1.6 present the same
relationship for another muscle directly involved in the active process of
assembling pieces. In Figure 1.5 the first hundred cycles at the beginning of
the shift give rise to about three oscillations. Including the initial practice
effect, it appears that the pattern of the subject's movements becomes
smoother, with a more economic exertion of the muscle groups involved.
This may be the reason for the decreasing electrical muscular activity in
combination with an increasing output rate. Figure 1.6 shows the reversed
behaviour of the time series of these stress and strain parameters for the same

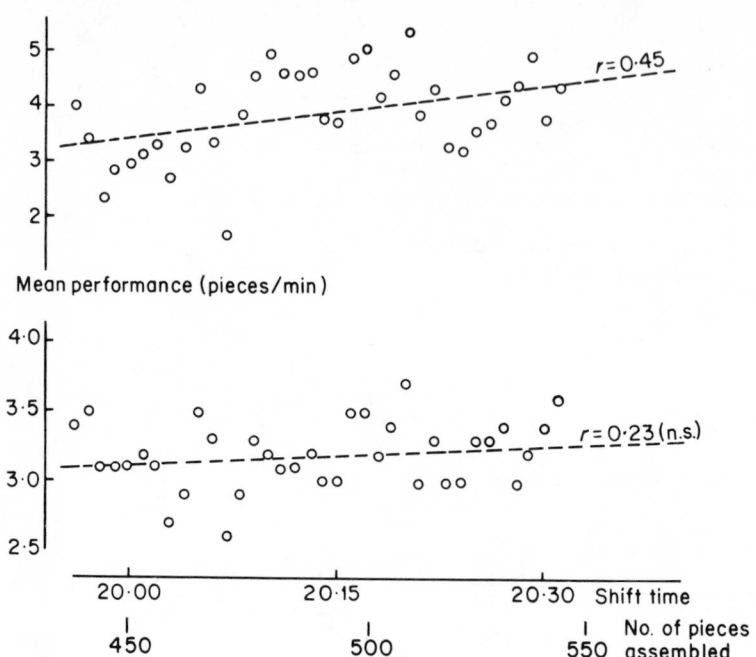

Figure 1.6   Electrical muscular activity and output rate at end of shift (M. pronator quadratus)

muscle group in the last hour of the investigated shift. In this case electrical muscular activity increases, whereas the output rate remains nearly constant.

In summary, these studies have shown an increasing electrical muscular activity in both the groups of muscles illustrated, which has to be interpreted as an increase in strain, arising as a possible result of a muscular fatiguing process. The conclusion is that the ergonomic design of the investigated working system could be improved if these indicators demonstrated changes implying that the increasing strain was being avoided.

This short discussion of the connection between stress and strain parameters illustrates the value of ergonomic field studies for devising new working structures and also for the comparison of current and prospective work arrangements.

over 180 beats a minute during intensive physical exercises. After an initial increase, the cardiac frequency is maintained at an increased level during the whole period of work. On cessation of the work the frequency starts decreasing, until it settles down at the prework level. The speed of the recovery will depend on the level of the activity (Figure 2.1) and the physical fitness of the person. The fitter the person, the quicker the recovery.

(b) Increase in systolic volume, i.e. the amount of blood that a single cardiac contraction injects into the circulatory system. The systolic volume may increase by over three times its resting value (about 60 ml during rest to 180 ml at heavy work).

(c) The increases in heart rate and systolic volume result in an increase in the minute volume, i.e. the amount of blood that the heart pumps into circulation in one minute. The minute volume is the product of the heart rate and systolic volume. It speeds up the blood flow, so that the active muscles get more oxygen while the metabolic products are removed more quickly from the muscles. The minute volume may vary, depending on the amount of work done, from four to 40 litres a minute.

(d) Increased use by the muscles of the oxygen supplied from the blood. While resting, muscles use only about 20% of the oxygen from blood. During increased activity, the oxygen intake by the active muscles is much higher (up to 80%), depending on the amount of work being done.

(e) Two other mechanisms that contribute to a better oxygen supply to active muscles are local vasodilation (in the muscles) and increase in blood pressure level. Blood pressure level may increase from about 130 torr during resting to over 200 torr during heavy physical work. It stays at a higher level throughout the work.

Changes in respiration due to work take place at the same time as the changes in the cardiovascular system. While the main function of the cardiovascular system is to supply oxygen to and take carbon dioxide from the active muscles, the main function of the respiratory system is to supply more oxygen to the blood and get rid of carbon dioxide during work. Measurements of oxygen and carbon dioxide exchange between the blood and the lungs have shown a direct relationship with intensity of work. The exchange is greatly helped by an increase in lung ventilation. Under resting conditions about half a litre of air enters the lungs by a single inhalation; the number of inhalations a minute is about 12, therefore only six litres of air enter the lungs in a minute. During physical activity lung ventilation increases as the product of depth (volume) and frequency of breathing. With very demanding physical tasks the volume of air entering the lungs per single inhalation may be over one litre, and the frequency may exceed 60 inhalations per minute. Under these conditions lung ventilation may reach the value of over 180 litres a minute.

Although oxygen consumption is directly related to amount of work, it is possible via the caloric equivalent of oxygen (1 l O_2 = 4.80 kcal) to express a

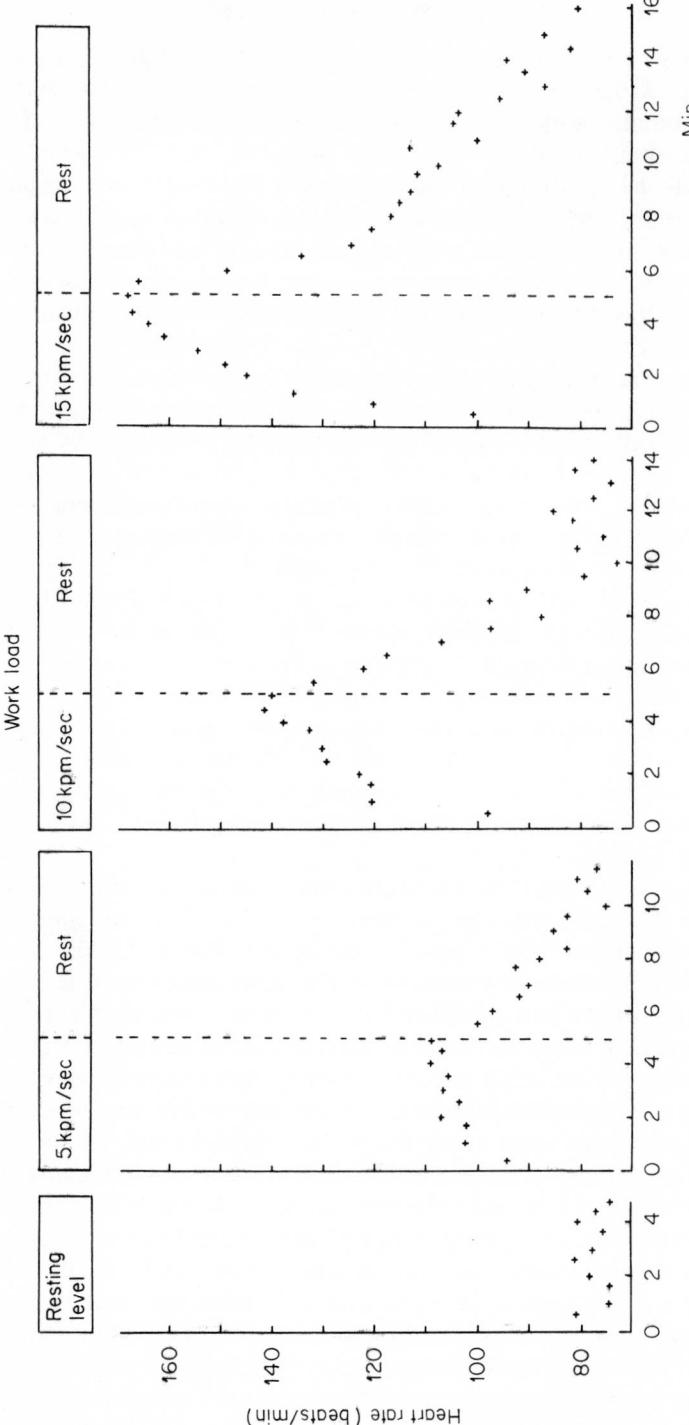

Figure 2.1   Changes in heart frequency during and after work of different intensities

given amount of work in terms of energy expenditure. This is an objective parameter of workload imposed on the operator not only by the work itself but also by other factors involved, such as work methods, workplace layout, environmental factors, etc.

Since both oxygen consumption and heart rate are directly related to amount of work, they also bear a linear relationship to each other within certain limits. In practice, it is easier to measure heart rate than oxygen consumption, which results in a greater use of the former parameter. If, however, the individual regression line between the two physiological variables is known (obtained *via* simple calibration tests), oxygen consumption may be worked out on the basis of heart rate (Tomlinson and Manenica, 1977, for example). By the use of this procedure interference with the work itself is avoided, especially if a telemetric recording of heart rate is used.

There are other physiological techniques for physical load assessment, but they have had a limited use in industry. Some of the major reasons for this are:

(a)   They measure changes in isolated bodily systems.
(b)   Good measurements are difficult to obtain.
(c)   They are rather expensive to use.
(d)   They require highly qualified personnel.

Work study vs. physiological indices

There have been relatively few studies involving a direct comparison between conventional work study and physiological work load indices. Results of the studies done by Wyndham *et al.* (1966) and Moores (1970) were far from conclusive. Morrison *et al.* (1965) found some correlation between work study and physiological indices of work load, which was most probably due to the fact that the variation in the workload was caused by the variation in the work rate. Generally speaking, where there was no change in the work rate but only in the effort involved, the physiological indices reflected the change while the work study did not.

Another study on the direct comparison of work study and physiological indices of work load was carried out by Tomlinson and Manenica (1977). The study involved 10 forestry workers, who carried out their everyday tasks at their usual speed. The tasks included felling and snedding, cross-cutting, work on sawbench, assembling and disassembling pallets, peeling, baling stakes, pushing bogies, stacking logs, and walking. The work study assessments were done by three experienced work study officers.

Physiological measurements included heart rate and oxygen consumption.

On the basis of these two variables combined physiological indices (CPI) were worked out for every operator doing different tasks. While the physiological recordings were being taken, work study officers were assessing the tasks carried out by the operators. The Spearman rank correlation coefficients between the CPIs and the work study ratings were obtained (Table 2.1).

Table 2.1   Correlation coefficients between CPIs and ratings. After Tomlinson and Manenica (1977)

Subject	Work study officer I	II	III	Overall
A	0.39	− 0.58	—	0.09
B	0.21	0.53	—	0.48
C	0.75	− 0.11	0.24	0.23
D	0.16	0.51	0.42	0.33
E	0.81	0.38	0.36	0.45
F	0.07	0.54	0.46	0.22
G	0.27	0.65	0.37	0.54
H	− 0.37	—	0.85	0.24
J	0.70	—	0.08	0.35
K	− 0.21	− 0.32	− 0.50	− 0.28

As can be seen from Table 2.1, there is little agreement between the CPIs and the ratings given by the work study officers. The correlation coefficients range from moderately high negative (subject K) to positive (subject E, for example). There is also little agreement amongst the raters. The ratings of rater I, for example, were negatively correlated with the CPIs in subject H, while the ratings of rater III had a high positive correlation with this index in the same subject.

On some occasions different work study officers assessed the same activities simultaneously and thus provided the basis for a comparison amongst themselves. Table 2.2 shows the Spearman rank correlation coefficients between their ratings.

Table 2.2   Correlation coefficients between the ratings of different raters. After Tomlinson and Manenica (1977)

Comparison	r_s	d.f.	Median constant error
Raters I and II	0.93	26	10
Raters II and III	0.88	19	0
Raters I and III	0.06	6	10

High correlation coefficients do not necessarily mean that the two raters made the same estimations of the tasks, as shown by the median constant error in the case of raters I and II. It is only in the second case (raters II and III) that the results indicate the existence of similarities in the assessment.

A complete summary of all the comparisons between the CPIs and the ratings of all the activities by the three raters demonstrated a rather poor correlation between the work study and physiological indices (Table 2.3).

Table 2.3   Overall correlation coefficients between the CPIs and all the ratings of the three raters. After Tomlinson and Manenica (1977)

Rater	r_s
I	0.46
II	0.08
III	− 0.12
Overall	0.13

The results showed that the work study ratings did not reflect the true workload imposed on the operator, revealing not only a great discrepancy between the two kinds of assessment but also an inconsistency amongst the work study specialists. While the physiological indices reflect the state of the operator, i.e. the effects of the work, the work study assessment is compounded with the variability resulting from the subjective factors of the assessor as well. Talking about the latter, Moores (1972) comments that the obvious method of reducing the effects of subjective factors in an assessment by the use of a number of assessors is in practice rarely carried out. Performance ratings are usually based on the assessment of a single work study officer.

There is, however, a whole range of work situations where an extra amount of workload is imposed on the operator by the postures required to carry out the work. This postural load is basically static work imposed on some muscles which resist the force of gravity and/or some other force affecting the body throughout longer work periods. Conventional work study completely fails to recognize this kind of work load. Physiological indices are somewhat better, but not sufficiently accurate in assessing it. In other words, static work does not affect the bodily systems in the same way as dynamic work does. Activity of the cardiovascular system in the active muscle resisting a constant force is impaired. The metabolic composition of such a muscle is different when it works at or above 20 % of its maximal force (Myhre and Andersen, 1971). As a result, it is not possible to relate accurately the changes in the cardiovascular and/or respiratory system to the amount of static work. Therefore, this is an area where both the work study and classical work physiology approaches to work measurement are inadequate.

A new approach

Instead of trying to measure work load objectively (physiological indices), or to assess it subjectively from outside (work study), an alternative is to try to assess it from the 'inside', i.e. through the operator himself. There are some encouraging results which indicate that this could be a successful approach to the assessment of static work load. Lloyd *et al.* (1970) showed that the subjective assessment of pain in the muscles involved in static work correlated with the EMG recorded on the same muscles.

In a series of experiments, Bujas (1972) combined subjective estimations of fatigue with some objective indices, such as EMG, heart rate, and residual work capacity. His results generally showed marked similarities between the subjective estimations and objective indices of fatigue or work load. The author summarized the experimental outcome as follows:

(a)   A subject is capable of making a judgement of his state of fatigue on a scale. The criteria of subjective evaluation appear to be reasonably stable and coherent.

(b)   The quantitative evaluations of fatigue show a reasonably good relationship with certain bodily changes during the work and periods of recuperation after the work.

(c)   It is possible, through a subjective evaluation of fatigue, to recognize well-known facts about the influence of work load, duration, type of work and training on fatigue.

(d)   Due to the methodological and other difficulties it is not yet possible to tell what the real relationship between subjective assessment and fatigue is.

(e)   Although there are no practical tests in this area which are sensitive and reliable, it may be said that the subjective evaluation of fatigue has some validity from the theoretical and especially from the practical point of view.

Corlett and Manenica (1978) tried to assess the contribution of static work load to general discomfort during the course of working time—both in the laboratory and in the industrial situation. Their results showed that discomfort caused by static (postural) work load increased with time, and it was possible to estimate it on a five-point scale. A further analysis showed a very good relationship between estimated discomfort and the proportions of posture-holding time in different situations.

In parallel with these investigations, Corlett *et al.* (1979) designed a fairly reliable technique for posture recording. The technique is easy to use, but its value in practical industrial situations is limited by the well-known shortcomings of subjective judgements carried out by human observers. The interpretation of the data is another, perhaps temporary, problem of the technique. For example, what does a bent body position of a certain number of degrees mean in terms of endurance, fatigue, discomfort, pain, etc.?

At this point, the only way out seems to be a symbiosis between different assessment approaches. That is, to use (and to improve) the posture-recording form and to search for the meaning of the 'deviations from the standard posture' in introspective reports and/or estimations from the operators. This could satisfy our short-term interest in the effects of postural load on the operator and his output. It would also be a good starting point for the study of long-term effects of postures on the operator (locomotor deformities, chronic pain, etc.).

In conclusion, it ought to be said that work study techniques are still needed in modern industry, but at the same time the industry does not need an outdated quasi-scientific approach to work measurement. The new approach should include the operator's self-assessment of the work load, as well as some other relevant variables. Such an approach should give the work norms based on present-day knowledge and the feedback from the operator. This would greatly improve this stagnant and controversial industrial service.

References

Bujas, Z. (1972). La validité des évaluations subjectives de la fatigue. *Le Travail Humain*, **35**, 193–204.

Corlett, E. N., Madeley, S. J., and Manenica, I. (1979). Posture targetting: A technique for recording working postures. *Ergonomics*, **22**, 357–66.

Corlett, E. N., and Manenica, I. (1978). The effects and measurements of working posture. In *Arbeidsplass og Miljø* (Ed. T. O. Kvalseth), Tapir, Trondheim, pp. 51–74.

Lloyd, A. J., Voor, J. H., and Thieman, T. J. (1970). Subjective and electromyographic assessment of isometric muscle contractions, *Ergonomics*, **13**, 685–91.

Moores, B. (1970). A comparison of work load using physiological and time study assessments, *Ergonomics*, **13**, 769–76.

Moores, B. (1972). Variability in concept of standard in the performance rating process, *Int. J. Prod. Res.*, **10**, 167–73.

Morrison, J. F., Brown, A., and Wyndham, C. H. (1965). A comparison of work study assessments and physiological measurements of men at work, *Trans S. African Mech. Engr.*, May, 1–5.

Myhre, K., and Andersen, L. K. (1971). Respiratory responses to static muscular work, *Resp. Physiol.*, **12**, 77–89.

Tomlinson, R. W., and Manenica, I. (1977). A study of physiological and work study indices of forestry work, *Applied Ergonomics*, **8**, 165–72.

Wyndham, C. H., Morrison, J. F., Williams, C. G., Hayns, A., Margo, E., Brown, A. N., and Astrup, J. (1966). The relationship between energy expenditure and performance index in the task of shovelling sand, *Ergonomics*, **9**, 371–8.

Stress, Work Design, and Productivity
Edited by E. N. Corlett and J. Richardson
© 1981 John Wiley & Sons Ltd

Chapter 3

Pain, Posture and Performance

E. N. Corlett
University of Nottingham, UK

Introduction

It is fundamental to the PROMSTRA approach that working methods should be devised which optimize the level of stress on the worker concerned. Hence, it could be argued, where some work load is excessive we may improve safety and performance by taking action to reduce it. But things do not always work out so neatly, and where posture and its effects are concerned a hypothetical case will serve to illustrate some of the problems.

Consider an operator engaged in a repetitive task who exerts high levels of effort and adopts some extreme postures during each cycle. At intervals he stops to rest, because his discomfort is too great to allow him to continue to work. A well-meaning methods engineer, recognizing the problem, modifies the work so that discomfort builds up over a much greater number of cycles, and points out to the operator the benefits of short working periods interspersed with short rest periods.

The operator, however, due to the improved work situation, is relieved of the rapid growth of discomfort previously experienced and works for longer periods. It appears likely that he works until the same level of discomfort is reached before stopping but, due to the changes, this means he can work a longer time and do many more cycles. The muscle groups involved in the work activities, although not so highly loaded at each cycle, are, nevertheless, used many more times per day. They are therefore more likely to adapt and cause the operator's body to grow into a shape more suited for the job he is doing.

That people's bodies adapt to their work is not a fanciful idea but has been shown to occur (see, e.g., Floyd and Ward, 1967). Whether it is better to experience extreme loads, with the risk of musculo-skeletal damage, or less extreme loads to which the body may adapt is a moot point. It must not be thought that any adaptation is bad, for it is a proper and necessary response to the environment. It is the extent of the adaptation which must be questioned, about which there are no firm rules and to which judgement must be applied. In the discussion which follows it is accepted that adaptation will occur, but the aim is to ensure that any changes do not lead to physiological asymmetry or disease.

It is well recognized that, for the cases described below, the changes proposed or introduced are not the best work situations which could be designed but are better, and equally feasible, situations than those which previously existed. Even better situations than those described are also perfectly feasible, but they require major redesign of the equipment itself before they can be realized; the results given below arise from modifications, in factories, to existing designs of equipment.

In what follows, then, there is no implicit support for the short-cycle repetitive task, merely a recognition that it exists and that many people earn a living by such activities. There is no evidence that tasks which constrain their operators to a limited range of postures are on the decline; indeed, they are being extended to a vast range of office jobs where workers, to do the job, must remain seated in front of screens or consoles. It is in this context that the study of posture must be viewed. Where there is any limit on a person's ability to adopt a posture at will, it is pertinent to ask if this constraint represents an unsatisfactory stress.

The criteria for satisfactory posture

In relation to posture, what can be described as 'satisfactory' and what is 'unsatisfactory'? To the ordinary citizen the postures adopted by a ballet dancer might be most unsatisfactory if he attempted to copy them, but the trained dancer remains active over an extended working life. The few operations performed by a seated press operator, on the other hand, may appear straightforward and harmless, yet their performance 10,000 times per day can lead to extreme levels of discomfort as well as considerable bodily distortion.

The difference between the two cases is not necessarily the level of 'fitness'; physiologically, each may be equally fit for the work he performs. The major difference is in the variability in posture which is possible in each job. In the case of the operator the same muscle groups are used every few seconds, and if the workplace design is poor then the attempt to use other muscle groups will inhibit his ability to do the task because he will have adopted a position from which he cannot do it.

To see the effects arising from a poor workplace design, consider variations on two factors relevant to performance. The first is the requirement to see the work. If the operator can sit upright and see from a range of positions, then changes in posture such as leaning forward or back or a little to one side will not hinder his performance on the job. These slight changes enable different groups of postural muscles to come into play. If, however, the operator must crouch, no matter how slightly, to see through a gap, then a change in posture prevents him seeing. To work, therefore, he must hold his trunk almost stationary in one position, giving no opportunity for relieving the small muscles of the back and neck which must be active to retain that position.

The second situation is where access to the work point is through a narrow space such as an open machine guard. If the width is such that the worker must hold in his elbows or lean sideways to gain access at all, then once more it is a limited number of muscle groups which are always called upon. If the width of the opening is such that the operator may have his elbows in *or* out, sit slightly to one side or askew and still gain ready access to the work point, then change is possible without hindrance to production and high performance is not accompanied by high discomfort.

Some common principles are now becoming visible as a result of this discussion. We are no longer considering 'one best way' but are looking at workplaces which provide a wide variety of ways, at the operator's choice. The effect of a workplace at the wrong height, a seat too high or low, a poor visual situation or any other of the many common mismatches between people and their workplaces is to restrict their choice of working positions and movements. It is true, and all too evident, that people will adapt to bad workplaces but it is also true that this is done at personal cost as well as resulting in lowered productivity.

Principles and practice

The previous section raised the question of criteria for deciding whether a given posture is good or bad, and the answer must obviously begin with 'It all depends . . .'. Even so, although rules and values cannot be specified for wholly defining the quality of a given posture, some principles can be enunciated for helping to make appropriate decisions during workplace design. Table 3.1 presents such principles; their development is discussed more extensively elsewhere (Corlett, 1978). Those principles higher up in the table take precedence in their application over those lower down in the sequence. These principles have two major applications in practice. The first is that, as already stated, they assist the person engaged in job design to decide which of the available choices before him is better than the others or guide him to seek for solutions which lie within the principles. Their second purpose is to provide a framework for thinking about 'man at work' which is realistic and represents a view of work as a totality which is as close to being accurate as it is feasible to achieve. The set of principles arises from a view of man at work in which a living person is recognized as handling equipment and information in a possibly unfavourable environment. It is not

Table 3.1   Principles for the arrangement of workplaces

1.   The worker should be able to maintain an upright and forward-facing posture during work.

2.   Where vision is a requirement of the task, the necessary work points must be adequately visible with the head and trunk upright or with just the head inclined slightly forward.

3.   All work activities should permit the worker to adopt several different, but equally healthy and safe postures without reducing capability to do the work.

4.   Work should be arranged so that it may be done, at the worker's choice, in either a seated or standing position. When seated the worker should be able to use the backrest of the chair, at will, without this necessitating a change of movements.

5.   The weight of the body, when standing, should be carried equally on both feet, and foot pedals designed accordingly.

6.   Work should not be performed consistently at or above the level of the heart; even the occasional performance where force is exerted above heart level should be avoided. Where light hand work must be performed above heart level, rests for the upper arms are a requirement.

7.   Rest pauses should allow for all loads experienced at work, including environmental and information loads and the time interval between successive rest periods.

8.   Work activities should be performed with the joints at about the mid-point of their range of movement. This applies particularly to the head, trunk, and upper limbs.

9.   Where muscular force has to be exerted, it should be by the largest appropriate muscle groups available and in a direction colinear with the limbs concerned.

10.   Where a force has to be exerted repeatedly, it should be possible to exert it with either of the arms, or either of the legs, without adjustment to the equipment.

11.   (Barnes No. 5) Momentum should be employed to assist the worker wherever possible, and it should be reduced to a minimum if it must be overcome by muscular effort.

12.   (Barnes No. 6) Continuous curved motions are preferable to straight-line motions involving sudden and sharp changes in direction.

13.   (Barnes No. 7) Ballistic movements are faster, easier and more accurate than restricted (fixation) or 'controlled' movements.

The following three principles have specific reference to movements in repetitive tasks.

14.   (Gilbreth No. 1 and No. 2) Both hands should preferably begin their therbligs simultaneously and finish at the same instant.

15.   (Gilbreth No. 4) Motion of arms should be in opposite and symmetrical directions instead of in the same direction and should be made simultaneously.

16.   (Gilbreth No. 13) To reduce fatigue, motions should be confined to the lowest possible classification as listed below, the least tiring and most economical being shown first.

       1st   Finger motions
       2nd   Finger and wrist motions
       3rd   Finger, wrist and lower arm motions
       4th   Finger, wrist and lower and upper arm motions
       5th   Wrist, lower and upper arm and body motions

a substitute for data, techniques or investigations, of course, but a framework to assist in judgements and decisions about work situations.

These principles provide guidance in design, or redesign, but measures of various factors are still needed if design is to be justified and its effectiveness demonstrated. In the following chapter P. Oja, I. Kuorinka and O. Karhu put forward a readily used method for evaluating observed postures. What the rest of this chapter will do is to show methods for using the effects of posture to decide where to introduce changes in workplaces as well as how to evaluate the effectiveness of the changes in terms of both the physical comfort of the worker and his performance on the job.

Pain as a measurement variable

In the introduction to this chapter it was explained how pain, if extreme, would limit performance, and it is intuitively reasonable that if pain rises above some arbitrary level the sufferer will cease the activities giving rise to the discomfort and seek some alleviation of it. The threshold will vary according to the individual: both his tolerance of pain and his ability to counteract it. Melzack (1973) has shown how people, by experience, can use body movements to inhibit the feeling of pain. At the same time, as people work at a job adaptations inevitably take place and their physiology and anatomy adjust to the conditions. It is therefore not surprising to hear a comment from a worker on the lines that, 'I found it uncomfortable at first but I've got used to it now and don't notice it any more'.

As stated earlier, because the body is a living organism which adjusts to its environment, adaptation will occur, and the changes which are noticed over the first few days of a new activity may well be considered acceptable. Where the same muscle groups are repeatedly used, however, and pain is evident even after considerable experience of the work, it is desirable that the job should be changed so that the muscles and joints can be relieved of their repeated loadings.

It may at this stage be questioned whether pain is a reliable enough measure on which to base design changes. There are now several studies which demonstrate the linear and regular growth of pain; an early one is that by Kirk and Sadoyama (1973). The experimenters found the maximum loads that each of their subjects could lift, with a suitcase in each hand, when their arms were extended 10 and 20° from their sides. Trials were then run in which a subject supported 30, 50 or 70 % of the maximum, with arms at 10 or 20°, for as long as possible. At intervals during the holding period he was asked to rate his level of pain on a five-point scale. Figure 3.1 shows the results of this study. The graphs have been 'standardized' by converting the values of holding time to percentages, enabling all results to be plotted on the same graph. It will be seen that, regardless of the loading, the growth of pain during the holding time was the same and that it was linear, half the holding time giving rise to half the maximum pain intensity in all cases.

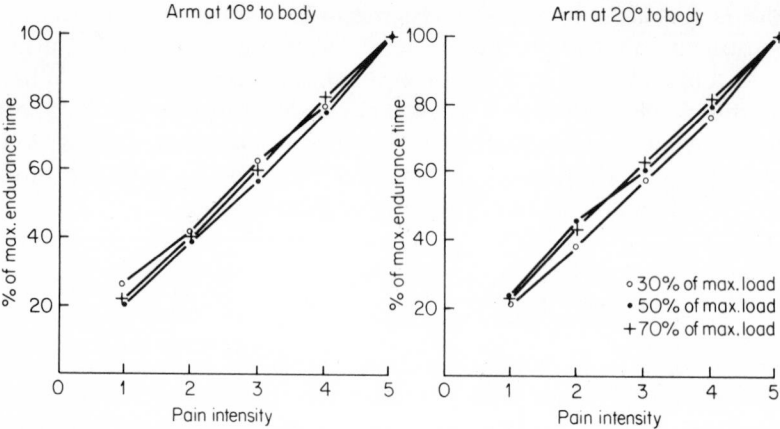

Figure 3.1   Appearance of various levels of pain intensity during static arm work. From Kirk and Sadoyama (1973)

Body pain and the body map

The study of pain given above asked of its subjects only an overall assessment of the level of pain, a question which is not specific enough if the more complex activities of work are to be studied. If there are several sites of pain in an operator, each at a different intensity, then it is probable that the most intense will be the key site which will trigger off the decision to stop work. However, apart from knowing the most extreme level, or obtaining some overall, integrated assessment of the total effect, it would be useful for design purposes to know the sites, and levels, of pain throughout the body. These would provide some guidance concerning likely causes of the pain as well as enabling before-and-after measures to show if improvements had taken place.

To help in redesign, the information in Table 3.2, after van Wely (1970), is useful. It arose from drawing together observations of symptoms in a large industrial surgery and of the jobs of those reporting the symptoms. The link between the two columns is not unequivocal in that on occasion some symptoms are reported which do not relate to the proposed cause, and some people who, according to the table, should be reporting the symptoms do not do so. Nevertheless, there was a very significant relationship reported by van Wely and the table provides useful guidance for changes which are of potential benefit to the operator.

Having a measurement scale (pain) which appears to behave consistently and some reasons (Table 3.2) for deciding on change, it remains to seek a technique for identifying the sites and intensities of pain. This is achieved by

Table 3.2   'Bad postures' versus probable sites of symptoms. From van Wely (1970)

Bad postures	Probable site of pain or other symptoms
Standing (and particularly a pigeon-footed stance)	Feet, lumbar region
Sitting without lumbar support	Lumbar region
Sitting without support for the back	Erector spinae muscles
Sitting without good footrests of the correct height	Knee, legs, and lumbar region
Sitting with elbows rested on a working surface which is too high	Trapezius, rhomboideus; and levator scapulae muscles
Upper arm hanging unsupported out of vertical	Shoulders, upper arms
Arms reaching upwards	Shoulders, upper arms
Head bent back	Cervical region
Trunk bent forward; stooping position	Lumbar region Erector spinae muscles
Lifting heavy weights with back bent forward	Lumbar region Erector spinae muscles
Any cramped position	The muscles involved
Maintenance of any joint in its extreme position	The joint involved

means of a body map and associated procedures (Corlett and Bishop, 1976). The map is shown in Figure 3.2. In use the words describing the parts represented by each division of the figure are omitted. The divisions are made according to the situation under study; for instance, for keyboard operators it might be useful to have a numbered section across each wrist. The numbers are for convenient reference by the person being questioned, and a range of maps with the areas differently numbered protects against the possibility of the memory of the number biasing subsequent trials.

In use the body map is employed together with a five- or seven-point scale, with the extremes 'anchored' by the terms 'no discomfort' and 'extreme discomfort', as some people see pain as something more specific than working discomforts. The first question asked of the operator is, 'Can you show me, by marking on this discomfort scale, what you judge to be your present level of overall discomfort'. When this general rating has been obtained, the body map is then brought out and the person is asked, 'Can you show me the part, or parts of your body which are, at the moment, most uncomfortable'. After the person has pointed them out, he is asked, 'Now,

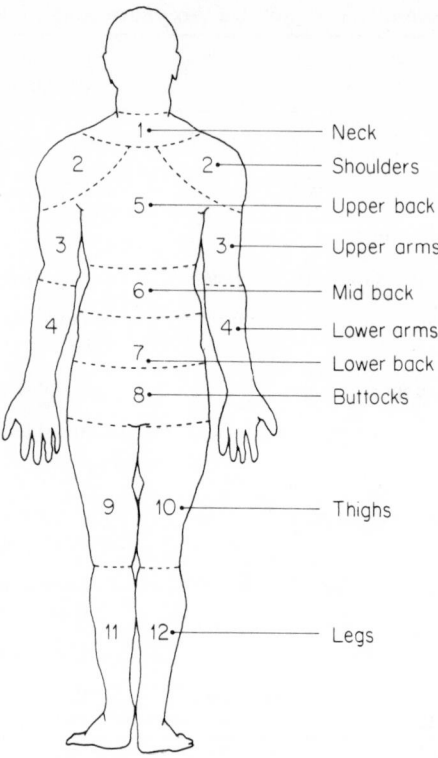

Figure 3.2   Body map. The regions into which the body is divided will depend on the postures and movements of the operator under study

which parts are the next most uncomfortable', and so on until no more parts are reported.

Interpreting the discomfort data

What the investigator has got the worker to do is to give an overall assessment of his working discomfort and also to point to all those parts showing noticeable discomfort. During this latter exploration the worker has differentiated between the various levels of discomfort he was able to distinguish. It is known that the judgement of sensations without an available reference standard typically results in four or five levels being distinguished, and when using this method at times when high levels of overall discomfort are reported, four levels of body part discomfort are usually and readily identified.

The procedure should be carried out at regular intervals throughout the working day to show the growth of discomfort as a result of the job. It should also be repeated on other days and with other operators. The reliability of the process is indicated by the closeness of results on different days with the same person. A few moments' thought will confirm that different people working similar machines can expect to show pain in different places since the postures they adopt will differ depending on their body sizes, methods of working, products being worked on, and so on. To combine results from different workers is probably not a good thing to do therefore, as they will tend to smooth out individual peaks and present a picture more favourable than that which really exists.

The results may conveniently be presented graphically. It will be found that body parts will often be identified by an operator in functional groups, e.g. shoulder, upper back, and upper arm. These may be combined during the analysis and a weighted score produced for the group as a whole. If the number of body parts combined in each analysis group is different, then a mean score, obtained by dividing the total by the number of parts in the group concerned, should be used.

The weighted score is obtained as follows. Each level of discomfort is weighted according to its intensity level, counting from the 'no discomfort' level as zero. 'No discomfort' is, of course, all those body parts which were not reported. The lowest reported level is scored 1, the next lowest 2, and so on up to the highest. As discomfort increases, so does the weighting value. The weights for each body part group are added together, divided by the number of parts in the group if necessary, and then plotted.

Posture records in practice

The records shown in Figure 3.3 are from a study by Mason (1975) and represent averaged values from four operators of pantograph engraving machines. When using these machines an operator guides a stylus with his right hand across a pattern which may be a set of letters. On a worktable is clamped the workpiece, and a high-speed rotary cutter, at the other end of a pantograph mechanism to the stylus, cuts the required letters larger or smaller than the pattern according to the setting of the pantograph. Operators of such machines may sit, although the pillar supporting the worktable causes a twisted posture, as they must set their knees to one side of it and may often be seen with their left hand on the cutting spindle bracket. This is partly for body support and can also provide some damping and aid control of the stylus. To get close enough to see the work, i.e. to achieve a viewing distance of some 300–400 mm, the operator must bend forward, and this brings the right hand and pattern plate to approximately shoulder level.

The effects of this on the operator can be seen in the figure. Even though these values are averages, and hence subject to some smoothing in the more

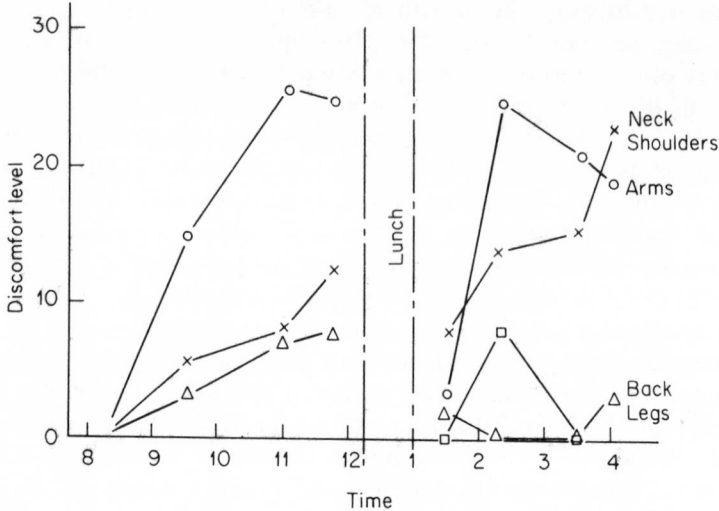

Figure 3.3   Levels of discomfort experienced by engravers, using pantograph engraving machines, during the working day

extreme cases, the discomfort levels reached by lunch time are high. Major discomforts arise in the upper part of the body, and it is very clear that the three-quarter hour lunch break does not lead to significant recovery. Both the arms and the neck and shoulders show rapid increases after lunch, restoring the curves to the levels reached prior to the break. In the case of the neck and shoulders group the climb of the curve then continues unabated.

On these machines there is little room for postural change and virtually no possibility at all of changing the position of the head, which must be within a certain range of distances from the component for the operator to see. The figure indicated that major attention should be paid to the position of the work in space in relation to the eyes of the operator when he is in a comfortable sitting position. The arm positions, particularly that of the right arm, should then be studied and the components of the machine rearranged until its technical functions could be achieved with the operator's arms and legs in more 'natural' positions, based for instance on the principles listed in Table 3.1.

Posture and performance

It will be recognized that some useful principles and methods exist for examining working posture and thereby evaluating the quality of a workplace, particularly in terms of comparing a workplace before and after a change to assess if the change has been of value to the worker. It has also been asserted that poor postures cause workers to stop work or reduce their output because the pain levels will not be tolerated. It is now time to examine this assertion.

A foot-operated spot welding machine requires its operator to hold the two components to be spot welded on top of a vertically mounted lower electrode whilst bringing the upper electrode down into contact with them by pressing a foot lever with one foot. The lever may have 250 mm or more of travel and, once contact is made, the further downward travel of the lever first increases the pressure between the electrodes and then switches on a low-voltage current. This passes from one electrode to another, through the two components. The resistance at the two contacting surfaces is enough, in conjunction with the high current available, to melt and fuse the two components together at a spot in line with the electrodes. The process is quick and, in certain circumstances, two or more spot welds a second may be achieved.

A study of these machines was made (Corlett and Bishop, 1978) with a view to improving the working conditions for users of existing spot welders, both operators and employers. Several machines were used, and event recorders attached to the machines recorded every stroke for at least two weeks before any change. Operators were asked for discomfort information by the techniques described earlier, the machines and operators were measured, and the postures adopted were recorded.

From these data the existing working conditions were evaluated. Daily output was recorded in the form of two histograms, one showing the distribution of the periods of time spent spot welding and one showing the time distribution of those periods when the machine was not working. It was hypothesized that, since the length of time to make one weld was controlled by the process, the result of improvements for the operator would be longer periods at the machine, since the discomfort of working would be less.

The working discomfort measures came directly from the body map investigation and were recorded in the manner of Figure 3.3. From a study of the working postures adopted the reasons for the discomforts could be identified and the relative importance of the different sources of discomfort judged. The posture was constrained by three machine aspects: (i) The need to stand on one foot (usually the left) and operate the foot lever with the other. Changing feet required changing the machine but would also have meant relearning the rapid hand–foot coordination needed to do the job quickly. (ii) The low level of the work point, set by the height of the tip of the lower electrode—a dimension which on some machines was not capable of being adjusted even to suit an average-height operator. This required a forward bend during the whole of each work period. (iii) Because the top electrode was set immediately above the lower electrode it obscured part of the component, particularly if a line of welds had to be made from front to back. The operator, therefore, usually leaned to the left to see around the top electrode.

For very small cost several machines were changed. Lower electrodes were set to appropriate heights, foot lever travel and force were each set to the minimum necessary to do the job, and, after experiments, foot levers positioned in relation to the lower electrode axis to minimize stretching to reach them. Finally, cranked top electrode tips (a standard and readily available item) were fitted so that the top electrode arm could be swung about 50 mm to one side of the centre line, allowing the operator to see the work more easily when producing a line of welds.

After the changes the measures were repeated, including having the event recorders running for at least two weeks. The results are shown in Figures 3.4, 3.5, and 3.6. Figure 3.4 shows the distribution of working and non-working periods for a foot-operated machine before and after alteration. The increase in the mean time of use is complemented by a decrease in the mean time for non-use, suggesting that the increased output is achieved with less effort, which consequently required shorter recovery periods.

This is supported by the graphs of Figure 3.5. The machine utilization and mean work and non-work comparisons show clear differences. The total discomfort score, whilst demonstrating clear improvement after the changes, does still represent a considerable level of discomfort. Discomfort was rated

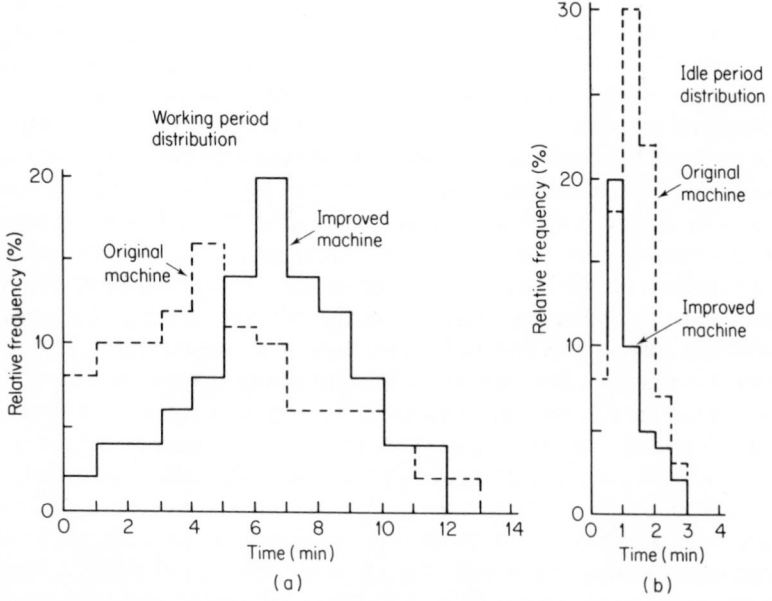

Figure 3.4   Changes in spot-welding performance after ergonomic changes to machines

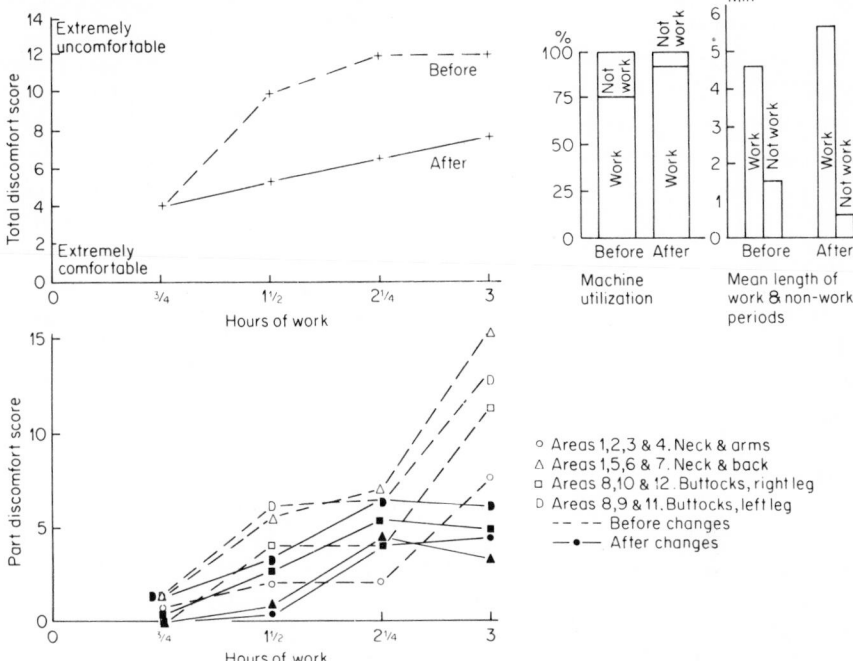

Figure 3.5   Machine 7. Foot-operated spot welder. Average of three operators

by subjects on a seven-point scale, which was then divided at each half-point to give the 14-point rating scale shown. The body parts score is the addition of the mean weighted score for each subject. Thus three subjects each having a mean weighted score of five would give a score of 15, as has occurred in this case for neck and back, implying extreme discomfort. It will be seen that neck and back, as well as the left (body supporting) leg, showed most discomfort and also show considerable improvement, particularly the former, which has moved from the top to joint lowest point on the graph.

Figure 3.6 shows the results for three power-operated machines. The only difference with these machines is that the electrode movement, etc., is done pneumatically and is initiated by a small foot control. This allows the operator to stand on both feet and to move the control around on the floor to find the best position to stand. Changes here only involved electrode height and the fitting of the cranked upper electrode tips.

Improved machine performance, whilst evident, is much smaller, as is change in the overall discomfort rating. (Note: the 'anchor' of 'extremely comfortable' printed at the bottom of these rating scales is no longer used, having been replaced by the term 'no discomfort'.) The part discomfort score

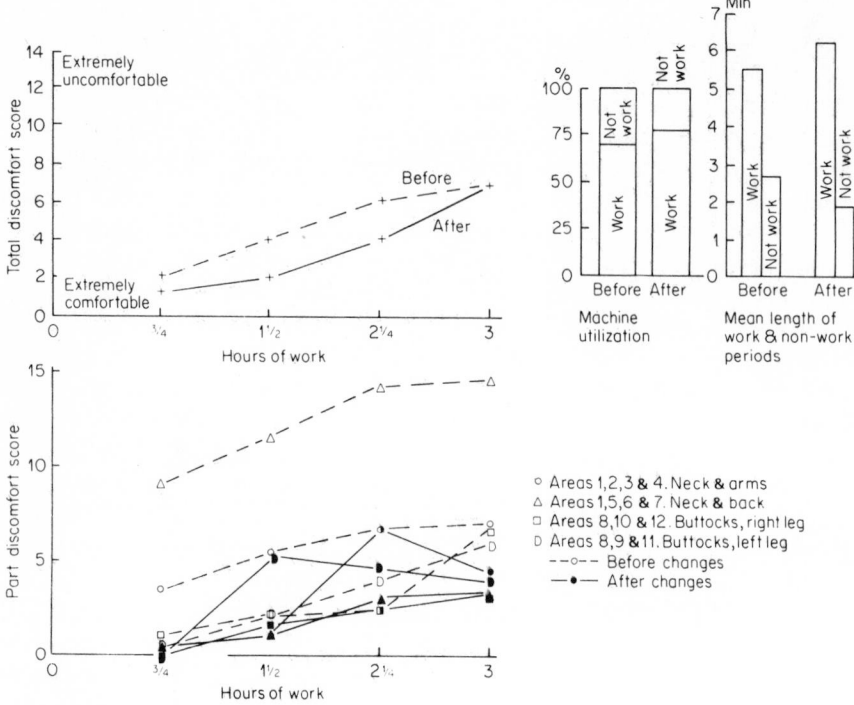

Figure 3.6   Machine 8. Power-operated spot welder. Average of three operators

shows a marked improvement for neck and back discomfort, commensurate with the changes introduced in working height and vision, but the low levels of discomfort arising from the foot control remain about the same.

It might be asked why the overall discomfort seen in Figure 3.6 shows so little change when neck and back discomfort is so dramatically reduced. It has been proposed that 'overall discomfort' is high when several body parts show significant discomfort, as in Figure 3.5, whereas if only one part is painful then people tend to interpret this not as general discomfort but, for example, as a pain in the back. Thus, if asked a question about their overall discomfort, they discount the single specific discomfort when responding.

This example describes a study of some machines where both human comfort and productivity were improved. The productivity improvement meant that not only did the operators earn their money more easily and more safely, but the improved machine utilization was equivalent to paying for the machines, just from the additional output, over a few weeks after the change. Although small in percentage terms, it was of major significance financially.

Some concluding comments

The spot-welder study demonstrated that there are gains to be made by both operators and management in the proper application of ergonomics to workplace design. There is, however, a larger context in which such changes are made, and the improved productivity will be crucially dependent on how the firm's organization supports the ergonomic activities. Piecework systems usually penalize operators who exceed what management has set as 'norms' of output, and the literature is replete with evidence of output restrictions to avoid rate-cutting. Hence the achievement of a successful job redesign in productivity terms depends on management rather than on operators. The rewards for working on the shop floor are not only financial; if the financial reward is denied, then increased control over the job is a valuable alternative. The scope for increased control may lie in an operator's opportunity to do the work at times to suit himself or to gain extra social interaction with his colleagues. Other advantages are possible depending on the opportunities available. It can be seen that an alert management will therefore consider its payment policy and clarify its objectives if it really wishes to gain full advantage from better working conditions.

Finally, let it not be assumed that matters of posture are minor and of little importance. Many of the diseases of middle age are contributed to in a major way by postural inadequacies. Low back pain, varicose veins, and tenosynovitis, to mention only three, affect large segments of the population and make their lives and leisure uncomfortable. There is no excuse for work-caused disease and it should only be tolerated if there really is no alternative.

It should also not be assumed that bad postures are experienced only by machine operators. The requirement to stand all day set by many jobs is unequivocally a bad posture. So, too, is the requirement to sit all day. Indeed, it is arguable that the latter is worse than the former. The problems of motor vehicle drivers are classic in terms of restricted posture and the problems of office workers are likely to follow them into the literature if some of the office automation on the horizon becomes widespread. We cannot afford to ignore posture, primarily because to do so creates such widespread misery, and secondly because the costs, both the social costs of unnecessary disease and the direct costs in lost productivity, are more than any modern industrial nation should be prepared to pay.

References

Corlett, E. N. (1978). *The Human Body at Work*. Management Services, May 1978.
Corlett, E. N., and Bishop, R. P. (1976). A technique for assessing postural discomfort. *Ergonomics*, **19**, 2, 175–82.
Corlett, E. N., and Bishop, R. P. (1978). The ergonomics of spot welders. *Applied Ergonomics*, **9**, 1, 23–32.

Floyd, W. F., and Ward, J. S. (1967). Posture in industry, *Int. Jnl. Prodn. Res.*, **5,** 213–24.

Kirk, N. S., and Sadoyama, T. (1973). A relationship between endurance and discomfort in static work. M.Sc. Report, Loughborough University, Dept. of Human Sciences, UK.

Mason, S. (1975). Measurement of comfort in industry. M.Sc. report, University of Birmingham, Dept. of Engineering Production, UK.

Melzack, R. (1973). *The Puzzle of Pain*. Penguin, London.

Van Wely, P. (1970). Design and disease. *Applied Ergonomics*, **1,** No. 5, 262–69.

Stress, Work Design, and Productivity
Edited by E. N. Corlett and J. Richardson

Chapter 4

A Method for Assessing Postural Stress in Industry

P. Oja and I. Kuorinka
Institute of Occupational Health, Helsinki, Finland
and
O. Karhu
Ovako Oy, Helsinki, Finland

Introduction

OWAS, or the Ovako Working Posture Analysis System, originated out of concern about the high prevalence of musculo-skeletal complications, especially low pack pain, among the employees of a Finnish steel and iron manufacturing company and their possible association with working postures. The method was developed as a tool for identifying those postures possibly responsible for the musculo-skeletal problems so that working conditions could be improved through the implementation of corrective measures and, eventually, through better technological planning for production.

The specific objectives of the OWAS project were the following:

(a) To develop a work study method for the identification of prevailing working postures in the company's steel and iron works;
(b) To evaluate the reliability of the method;
(c) To classify the postures according to the subjectively evaluated strain on the workers;
(d) To construct a procedure for determining the necessity of measures based on posture analysis and the classification of the effects.

Development of the system

Classification of postures

First, 700 jobs were photographed so that the postures used in the plants

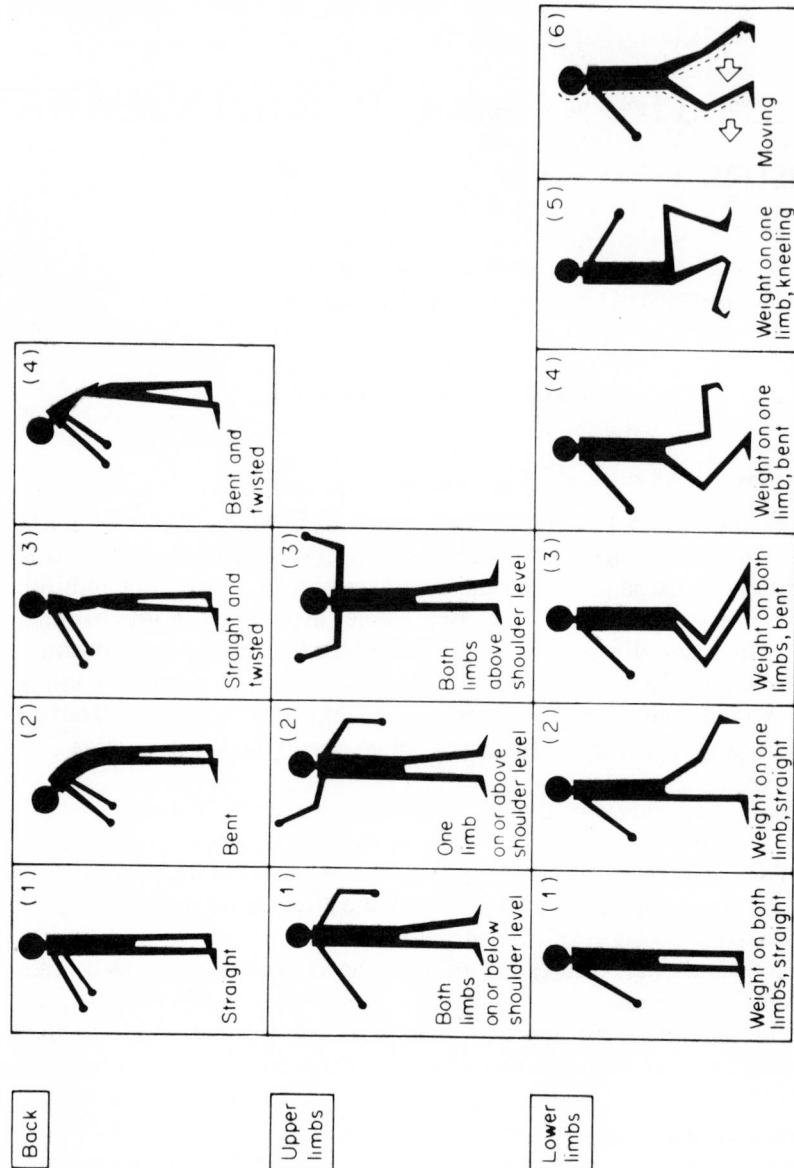

Figure 4.1   The basic postures of OWAS for the back and upper and lower extremities

could be surveyed and described. The main elements of the postures were constructed on the basis of the following coded positions of the back, upper limbs, and lower limbs (illustrated in Figure 4.1):

Back: 1. Straight
                2. Bent
                3. Straight and twisted
                4. Bent and twisted
Upper limbs: 1. Both limbs at or below shoulder level
                2. One limb at or above shoulder level
                3. Both limbs above shoulder level
Lower limbs: 1. Weight on both limbs, straight
                2. Weight on one limb, straight
                3. Weight on both limbs, bent
                4. Weight on one limb, bent
                5. Weight on one limb, kneeling
                6. Moving

In addition to the positions of these three parts of the body, the load being handled—either its weight or the force applied—was classified as (a) less than 10 kg, (b) 10–20 kg, and (c) more than 20 kg. A given posture was defined through the assignment of one of the classified positions for the back, arms, and legs. A four-digit code was obtained, the first three numbers identifying the position of the back, arms, and legs respectively and the fourth the load being handled. Thus 2132 means: back bent, both arms below shoulder level, weight on both lower limbs bent, load to be handled around 10 kg. With four possible positions for the back, three for the arms, and six for the legs, the total number of possible postures was 72.

Reliability of observations

The posture classification was used in the classical work-sampling fashion, i.e. split-second observations were made at regular intervals during the workday, and the occurrence of different postures in the plants was determined. Data were collected by six work study engineers on two different occasions; they comprised a total of approximately 36,000 observations.

The reliability of the method was evaluated from results of the second observation period. Interobserver reliability for 12 jobs was determined between six observers. The mean agreement was 93% with a range of 74–99%. The mean agreement for interworker reliability in a given job was 69% (23–88%), and the stability of observations between morning and afternoon 86% (70–100%). The interobserver reliability was considered to be satisfactory. The considerably lower interworker reliability suggested that the way different individuals do a given job can vary considerably. The morning *vs.* afternoon differences for a given job were generally not very great.

Evaluation of postural strain

The next phase of the project was to evaluate the musculo-skeletal strain of the classified postures. Both workers and experts classified all 72 postures according to the subjectively judged degree of strain and discomfort they caused.

For the worker evaluation 29 men, all with a minimum of five years' experience in the factory, were selected. They formed the panel for assessing the strain and discomfort of the postures. Using a four-point scale, they classified the complete set of postures twice in a randomized order. The point scale for the classification was defined as follows:

1 = Posture is experienced as normal
2 = Posture causes a normal feeling of fatigue after the workday
3 = Posture is experienced as particularly straining during, and the worker is unusually tired after, the workday
4 = Posture is experienced as causing pain and tiredness even after only a short period

The expert evaluation was made by six experienced ergonomists from five European countries. They used the following six-point scale:

1 = Very mild physical strain
2 = Mild physical strain
3 = Almost severe but still mild strain
4 = Almost mild but rather severe strain
5 = Severe strain
6 = Very severe strain

All 72 postures were evaluated at three load levels, i.e. no load, 10 kg on average, and 20 kg or more on average. Part of the ranking order of postures with a 10 kg load, as evaluated by the workers, is shown in Table 4.1.

In the final classification, the workers' rating, weighted by the experts' evaluation, formed the basis for classifying the postures into four categories. This classification for the 72 postures with a 10 kg load is presented as an example in Table 4.2.

It should be noted that the duration of any given posture, its frequency, and the particular effects of any sequence of several postures are not taken

Table 4.1 Part of the ranking order of postures as evaluated by the workers

Posture Code	Description	Ranking score[a]	
112	Back: straight Arms: both below shoulder level Legs: one leg on the ground, straight	− 2.55	Least straining
111	Back: straight Arms: both below shoulder level Legs: both on the ground	− 2.31	
—			
—			
—			
412	Back: rotated, bent Arms: both below shoulder level Legs: one on the ground, straight	− 0.16	
222	Back: bent Arms: one above or at shoulder level Legs: one on the ground, straight	0.00	
—			
—			
425	Back: rotated, bent Arms: one above or at shoulder level Legs: one knee on the ground	1.73	
435	Back: rotated, bent Arms: both above or at shoulder level Legs: one knee on the ground	1.75	Most straining

[a] Based on Torgensen's Law of Categorial Judgement (Björkman and Ehman, 1962).

Table 4.2 Classification of the postures into four categories based on the strain caused by them. The postures are identified by their code numbers

Category 1		Category 2	Category 3	Category 4	
112	126	134	215	421	433
111	114	124	135	125	224
122	316	326	213	426	324
312	212	431	223	336	415
132	133	412	226	333	234
131	221	222	231	314	334
116	123	413	331	423	325
216	311	136	422	315	235
113	411	232	233	225	434
322	332	321	323	414	424
121	115	432	416	335	425
211	214	313	236	436	435

into consideration in the evaluation, even though they are important elements as regards the effects of the postures.

Indication for corrective measures

The final stage in the development of the system was to define a standard procedure for collecting observations and for taking corrective measures when necessary. The four strain categories were rated according to the urgency of implementing corrective measures, as indicated in Table 4.3.

Table 4.3   Urgency of implementing corrective measures for the four categories of postures

Category	Degree of strain	Urgency ratings
1	Postures as such are not harmful	No particular measure necessary
2	Work involves postures which have significant strain effects	Corrective solution implemented in the near future
3	Work involves postures which have very significant strain effects	Corrective solution implemented as soon as possible
4	Work involves postures which have obvious harmful effects	Corrective solution implemented at once

For the collection of posture data a summary sheet similar to the one presented in Figure 4.2 was designed, the only difference being that the four urgency categories were colour-coded for easy evaluation of the results.

Application of the method

When OWAS is applied as a work study tool, data are first collected in a work-sampling fashion. The working postures are observed at regular intervals and classified as one of the 72 possibilities based on the momentary position of the back, the arms, and the legs. The observations are recorded on a data sheet which has a specific space for each position (Figure 4.2).

Figure 4.2 shows the summary of observations made during the bricklaying of the deck of an electrical arc furnace. Of the 239 'working' observations, 76 are in an additional class 7 for the lower extremities; this class indicates special trunk movements. Fourteen other observations fall into categories 3 and 4, because the worker's back was bent and/or twisted. After the collection of the basic data an ergonomic analysis of the same job is made with special emphasis on the critical back positions.

Worker		Duration min	No. of observations 385
Research worker		Observation min, interval min	

	back	upper extremi-ties	lower extremities – use of strength																				
			1			2			3			4			5			6			7		
			1	2	3	1	2	3	1	2	3	1	2	3	1	2	3	1	2	3	1	2	3
	1	1	69		1		9													3	65	5	1
		2																			3		
		3																					
	2	1	31			3	3		5			8			2								
		2					2		1	1													
Total number of observations at work 239 % min		3						1		2													
	3	1		1					1	1											1		
		2																					
		3																					
	4	1	10		1	1			1	3	1				1						1		
		2	1																				
		3																					

	Pause	Waiting	No observation
Not working	87	56	3

Figure 4.2 The data sheet for observations. A summary of a work study on bricklaying is illustrated. The urgency category for each posture can be determined from the different shades of grey. The darkest grey represents category 1 and no shading at all category 4 (see Table 4.3)

An example of the application of the OWAS method in a steel mill factory

The deck (lid) of an electric arc furnace is a calot-like construction made of firebrick. The deck wears rapidly and must be rebuilt weekly.

In the old working method the deck, having a diameter of 5.3 m, was built on site from heavy fire-resistant bricks, layer upon layer. The tasks were physically heavy, and many poor postures were observed.

An ergonomic group of four members (the foreman, an industrial engineer, and two workers) was appointed to solve the problem. The task began with an OWAS analysis of the working postures involved in bricklaying.

The bricklaying was done primarily in a bent posture (43% of the working time). The work was the most strenuous during the laying of the first row of bricks, when the bricklayer had to work in a deeply bent posture.

After the posture analysis the group went on to consider technical innovation and design. Their solution was an element bricklaying system in which a prefabricated element was built in better, more controlled positions. The assembly was then accomplished with a crane.

After the technical changes a new posture analysis was made. The time spent in a bent posture decreased. The new system has been in use for more than two years and has proved to be less strenuous on and more acceptable to the workers.

Discussion

The need to study working postures in an industrial setting is usually based on three premises:

— Poor postures distort the normal way of working or cause tiredness or pain
— Poor postures are connected with work accidents
— Poor postures are harmful to health.

The role and relative importance of each item, i.e. ineffectiveness or tiredness, accidents and harmfulness to health, are not easily distinguished from each other. With the OWAS method discomfort is the main criterion for poor posture.

The need to analyse postures at the workshop level arises in many industrial functions. Health officers are interested in postures as causes of musculo-skeletal diseases, while safety engineers regard bad postures as possible causes of accidents. Poor postures as causes of unnecessary human stress, and thus as causes of inefficiency, interest the work study engineer.

OWAS was designed to be used by the work study engineer, either as a part of his daily routine or as a supplementary aid. Therefore the procedure (work sampling) was directly linked to the standard method used by Finnish work study engineers. The ultimate goal of the method was not, however, job efficiency as such. Instead it was health-oriented and was achieved through the selection of classes of postures known to be related to causes of back and upper limb diseases in workers. Improved efficiency is expected as a secondary result as postural discomfort and possible causes of back–upper limb afflictions diminish.

In Finland the use of OWAS requires a certain amount of basic training in work study methods as well as additional training with the system itself. The training is supervised by the labour market organizations.

OWAS is meant to be used in an enterprise as part of the daily work study routine. Experience with the system under such conditions has already been collected in a steel company over a two-year period. The results show that the analysis and the method are not the most important aspects of the study of working postures, but rather the operational integration involved. Analysis without corrective measures and corrections without input into the planning of production technology lead to activity where the output is small in relation to the effort expended. In the company in which OWAS was developed, a balance has been achieved between these factors, and consequently the improvement of working postures is not an isolated activity but, instead, part of the developments in working conditions that are taking place through planning, production, and labour protection.

Reference

Björkman, M., and Ehman, G. (1962). *Experimentalpsykologiska Metoder*. Almqvist & Wiksell, Stockholm, p. 156.

PART II

Methods of Assessing Psychosocial and Mental Stress in Industry

Introduction

All jobs involve some sort of mental activity and the contribution of this mental activity to overall workload is an important area of scientific investigation. Enormous advances in industrial technology and the widespread introduction of data processing systems in industry have resulted in important changes in the task content of the lower-skilled jobs. Physical activity is playing a lesser part in most jobs while cognitive activities are becoming more and more important. Production workers are less involved with manual transfer, lifting and machine operating tasks and are being asked to monitor automatic machines or processes, consult data processing systems, and generally be responsible for taking more important decisions. The consequences of these changes are discussed in Part III (see e.g. Eason and Sell), and often result in 'taking the interest' out of the task while maintaining the repetitive routine activities. In monitoring tasks, the normal task conditions can be considered as underloading the operator; however, the work can be punctuated by periods of hyperactivity, during a breakdown in the system for example, where the operator is overloaded. The difficult problem for the system designer is to know how to design the production system so that the operator is not overloaded beyond his capacity when attempting to recuperate a breakdown.

Two recent surveys of research work in mental load demonstrate the interest that is being shown in this area. The NATO Special Panel on Human Factors ran a conference on the theory and measurement of mental workload (Moray, 1979), while the International Congress of Psychology of 1976 also devoted a special session to mental workload (*Ergonomics*, 1978). Both these surveys have attempted to put forward a general methodology for the assessment of mental load. Mental load is considered as a general construct and any assessment of mental effort includes measurements of task demands, operator effort, and performance (Johannsen, 1979). To understand operator effort it is necessary to know the demands made by the

55

task, to study his behaviour while carrying out the task, and to have an idea of the costs of achieving a reasonable task performance. The task demands can be determined from the workplace and task organization. These variables are usually stable although they can vary according to different production schedules. The measures of operator effort pose considerable problems because the processes which are being considered are internal. The main measures used fall into three categories: information processing measures, physiological measures, and subjective effort ratings.

Part II includes two chapters which review two types of measure. However, the chapter by Kalsbeek also includes proposals which give a view of mental load as a combination of the accessing of data for the production of behaviour and its selection in relation to the demands on the person. The load arising from mental activity includes the necessity to gain access to the appropriate subsets of data for the task in hand. A greater or lesser threshold may have to be overcome to achieve this, according to internal and external circumstances. This threshold may be spontaneously overcome, or can be influenced by training or deliberate recall by the individual. Kalsbeek briefly indicates other aspects of load too, and why variations in levels of this load during a working period can be advantageous.

There is still the underlying assumption that information processing capacity in man is restricted by a single channel flow of information which is of limited capacity. These theoretical restrictions have been criticized and it is clear that care must be taken in the use of measures incorporating this assumption. However, secondary task methods have been successfully employed in very controlled situations although their use on the shop floor presents many difficulties (Brown, 1978; Pew, 1979).

The physiological measures of operator effort are reviewed by Strasser. Changes in mental activity may well influence physiological processes not directly associated with brain functioning. These processes are much more accessible for measurement than cerebral processes. It is hypothesized that the changes in certain physiological parameters are globally related to operator mental effort. Strasser discusses some of the problems associated with the recording and interpretation of physiological variables, especially cardiac arrythmia. Cardiac or sinus arrythmia is the variability of the beat-to-beat intervals of the heart. At rest, the heart beats slowly and the interval between the beats is variable. As physical exercise increases, the intervals between heart beats become more regular. The variability between intervals also changes with increased mental effort, without necessarily an increase in heart rate. This change in variability has been used to compare mental effort for different types of tasks.

Laboratory studies of physiological measures in mental work do not ask people to do tasks which represent industrial jobs; people do not do mental arithmetic or solve anagrams all day. The use of a single measure in an industrial task is not advisable; it is preferable to monitor a number of variables during a task. The robustness of each variable cannot be relied on during field recordings as the external influences from the work environment are not so closely controlled as in the laboratory. There is a tendency in field studies to monitor as many variables as possible in order to detect whether there are concomitant effects.

The second half of Part II is concerned with occupational stress. Cox and Mackay summarize the present theories of occupational stress and propose a theory based on the idea that a person develops a transaction between his mental state and his social and psychological environment which enables him to cope with the different demands made on him. The next two chapters describe studies of the incidence of stress symptoms in two different working populations. Marshall and Cooper studied the various job factors in different managerial departments which contributed to stress symptoms. Tinning and Spry discuss the incidence of mental health problems in the steel industry and have related these to changes in the organizational structure of the industry.

Cox and Mackay attempt to clarify the problem of defining occupational stress in the introduction of their contribution. The use of the word 'stress' has led to a number of confusions in the literature. It has been used in an engineering sense, stress being understood as a measure of the applied condition which causes increased operator demands and strain as the measure of the effects of the stress on the operator. The other common use of stress has been derived from the work of Hans Selye, where 'stress' is considered as the general syndrome of effects from 'stressors' applied to the individual. The concept of 'occupational stress' is relatively recent and its popular use in the media demands that a precise definition of the term be given when it is used in a scientific paper.

Numerous studies have attempted to isolate the stressing factors in different occupations and to assess the extent of the effects of these stresses on different populations. The most commonly studied populations belong to the higher-level employment categories such as managers, doctors or pilots. However, it is becoming evident that occupational strain symptoms or mental health problems associated with work are common to all employment categories. The extent and significance of mental health problems throughout industry are not well documented and little is known of the contribution of working conditions to the breakdown of mental health.

The two field studies of occupational stress concern two different populations. Marshall and Cooper study the factors which can be considered as managerial stresses in different managerial jobs. The study initially looked at stresses resulting from the relocation of managers to different sites within a company. The conclusions from the first part of the study demonstrated that mobility was not the primary problem but highlighted the problems which are inherent in managerial jobs. The second part of the study compares the stresses present in different managerial departments. Marshall and Cooper also discuss the methodology employed in the study and indicate the advantages and disadvantages of using statistical techniques to analyse questionnaire and interview data.

In the second case study Tinning and Spry report an investigation on the effects of certain changes in work organization on the mental health of the workers involved. The study extended over many departments with the objective of identifying the trends and symptoms of undue mental strain. Previously diagnosed long-term cases of mental ill-health were not included in the results.

Departments undergoing reorganization or under a threat of redundancy among the workforce showed serious levels of mental health deterioration, about 5% of the workers demonstrating that they were in need of help or some treatment. The results from departments which were not undergoing the same sort of changes showed a definite difference. The alarming factor in this study is not that certain people were found to be suffering from a severe deterioration in mental health but that a very large group showed a significant deterioration. This group is probably made up of people whose personality does not predispose them to trouble, but their health has been affected by the circumstances in which they find themselves. These are the people who could benefit from counselling or advice on how to deal with their present situation. If they can be helped to cope through a difficult period they will probably profit from the experience of successfully dealing with it and subsequently will be better prepared to face other difficult situations. Since there is presumably no advantage to either the workers themselves or the company in exposing people to stresses which reduce their mental state to some level below 'normal', the study also underlines the need for a more ergonomic approach to change, in which situations are devised to enable people to cope best with the environment to which they are exposed.

References

Brown, I. D. (1978). Dual task methods of assessing work load. Ergonomics, **21**, 3, 221–4.

Ergonomics (1978). Symposium on Mental Work Load. *Ergonomics*, **21**, 3, 141–234.

Johannsen, G. (1979). Workload and workload measurement. In Moray, N. *Mental Workload, its Theory and Measurement*. Series II: Human Factors. Plenum, New York.

Moray, N. (1979). *Mental Workload, its Theory and Measurement*. New York NATO Conference series. Series II: Human Factors. Plenum, New York.

Pew, R. W. (1979). Secondary tasks and workload measurement. In Moray, N. *Mental Workload, its Theory and Measurement*. Series II: Human Factors. Plenum, New York.

Stress, Work Design, and Productivity
Edited by E. N. Corlett and J. Richardson
© 1981 John Wiley & Sons Ltd

Chapter 5

The Production of Behaviour and its Accompanying Stresses

J. W. H. Kalsbeek
University of Twente, Enschede, Netherlands

Introduction

With increasing mechanization and automation mental workload becomes a factor of crucial importance, both for production and for health. There are many concepts of mental workload but in this chapter it is defined as the biological impact of the voluntary control of the production of behaviour. It is considered that a classical behaviouristic approach comparing information inputs with task performance is not an adequate definition for the purpose of examining mental load. A human being is conceived of as a behaviour production system. As long as a person is alive this production will go on. The control of this production process is partly involuntary and partly voluntary. Voluntary control can be exercised in a more or less economical way with regard to the consumption of 'mental energy'. The general production procedure is that sequences of programmes for elements of behaviour are selected which, when triggered, result in the output of the observed behaviour.

Mental activity is defined as the acquisition and manipulation of data elements (programmes, data, data structures). Constraints to the sequencing process partly arise from the need to protect the dynamic integrity of the internal world (*milieu interne*) and are partly imposed by the external situation. An example of such a situation is a man–machine system. At the input interface of this system information is presented to the biological component. At the output interface action is imposed upon the machine component. Information inputs are supposed to guide rather than cause behaviour to be produced.

In the detailed analysis of behaviour production which follows, aims, constraints, procedures, functions, levels of organization and means are distinguished in accord with the description of artificial systems given in

Chapter 11. Relative contributions of each step in the production of behaviour are related to the overall mental load, whose contribution to the level of health and well-being is evaluated.

In the model described here the concepts of mental energy and mental capacity are used in analogy with supply and consumption of electrical energy in, for example, communication systems, but in our case without specifying the nature of the energy source or the biological structure of the capacity.

Mental workload and work physiology

Work physiology as a medical discipline is concerned with the problem of mental workload only when an individual experiences negative effects due to controlling his own behaviour: negative effects in the sense that his socio-psycho-biological system functions less than optimally according to agreed criteria of health and well-being. Thus mental workload has to be seen in relation to individual health. Social problems and feelings such as work satisfaction are again only of concern to work physiology insofar as there is a relationship to individual health.

The problem of mental load in relation to health can be defined as the difference between the state of an individual's socio-psycho-biological system at time t_1 and the state of it at time t_2, after a period of controlling the shaping of behaviour. If such a difference can be demonstrated, it is to be evaluated in terms of the risk to individual health (as has been said above).

An individual's socio-psycho-biological system can be conceived of as a behaviour production system which operates throughout his life. This production can be controlled at different levels of consciousness ranging from reflexes to the deliberate performance of chosen behaviour patterns. Production methods can differ, for instance, with regard to the intensity of control. Also, the factors which contribute to the production of behaviour such as physical effort, information processing, and emotions can be of variable importance. These three components act in an intrinsically interactive manner.

Behaviour as a product can be used to serve different purposes in different settings. Such settings are family life, leisure occupations, political and religious groupings, holidays, and, of course, man–machine systems. In all these cases individual behaviour as a product is used either to continue, discontinue or shape those settings. The principles of behaviour production remain the same regardless of the settings and purposes.

In engineering production, systems are defined by objectives, procedures, and functions. The functions are to be fulfilled by various processes and means. Objectives are to be reached within defined production constraints such as quality, cost, time, ergonomics, risk or pollution. An attempt will be made below to describe individual behaviour production in a way analogous to the production of engineering goods. It will be argued that the general procedure of behaviour production can be defined as a sequencing of behaviour segments arising from sequentially triggered control programmes. The selection process to trigger these programmes will exhibit a hierarchical structure. Behaviour production involves performing such functions as perceiving, interpreting, choosing, etc. The general procedure remains the same regardless of the purpose for which the behaviour is produced, i.e. whether there is spontaneous behaviour production or whether the behaviour being produced has constraints imposed such as man–machine systems. The functions also remain the same regardless of the relative importance of each function in producing the required behaviour. The physiological systems which bring about the various functions also remain invariant.

Goals in behaviour production

It took millions of years for the human behaviour production system to attain its present state. Industrial and other more sophisticated man–machine systems arose only recently. So system designers should not proceed as if human systems were especially fit for functions in man–machine systems, waiting to be programmed with regard to industrial objectives and constraints.

The human behavioural system was developed over millions of years in interaction with a natural environment. Time constants were large and variations in the environment normally highly predictable, except in attacking or being attacked, when the information flow would be high. Although uncertainty was low, the danger level was generally high. With the introduction of man–machine systems a new and competing system to the natural one has arisen. The objectives of spontaneous behaviour production and behaviour production under the constraints of a man–machine system can conflict, and a person can be forced continually to choose, as he can only give priority to the conflicting requirements one at a time.

Functions of human behaviour production

Behaviour production generally is guided by perception. Perception informs the person about the state of something inside or outside himself. This information about the actual state is compared with the state which would be preferable according to certain criteria and leads to the choice of a behaviour pattern intended to change the actual state into the desirable one. What is actually perceived is often not enough to permit a correct comparison. The incoming information has to be interpreted, i.e. to be 'enriched', by adding information from a mental image of the situation. A mental image is the sum total of an individual's knowledge about how a situation is structured and works. It can, however, be (partly) erroneous and has not necessarily to be consistent—the term 'image' can be misleading. Rightly or wrongly, an individual uses his own mental image as a reference in controlling his behaviour production.

When a behaviour pattern is chosen, the corresponding programmes which are necessary to produce this pattern have to be triggered in sequence.

So far the following functions of the human system for behaviour production have been distinguished: perception, information enrichment, comparison (actual control), choosing, sequential triggering of control programmes, realization of programmes, and giving priority to objectives and constraints from different sources. Also, three different levels of control have been distinguished:

1. Physiological subsystems acting as supplying systems;
2. Functions in behaviour production;
3. Priority regulation in cases of conflict about objectives and constraints derived from different uses of the behaviour as an end product.

Biological functioning and risk to health

Functions are usually defined without specifying the means by which they are actually fulfilled. Functions of human behaviour production are fulfilled by the activities incurred by the person's perceiving, choosing, etc.

These activities belong to an integrated, individual, (psycho-) biological system, the *milieu interne*, which defends its integrity by maintaining variables between tolerance limits. These variables are assumed to be correlated with the intensity at which physiological subsystems are functioning during a certain time. Figure 5.1 shows the hypothetical relation between intensity of functioning and risk to health. Physiological subsystems are taken to function around a setpoint within broad tolerance limits. Varying the intensity of functioning between these limits does not change the risk to health. Outside the limits the risk increases according to an exponential law. Functioning at higher levels is called overload, whereas functioning at lower levels is called underload.

Functioning at a certain degree of intensity is supposed to be a condition for maintaining the subsystem in a healthy state. As is well known, physical underload for long periods leads to a loss of function, e.g. atrophy of muscles. Long periods of mental underload can cause a shift back to lower levels of vigilance. Overload implies the use of reserve capacity. For shorter periods this use is acceptable, but for longer spells it leads to exhaustion. This

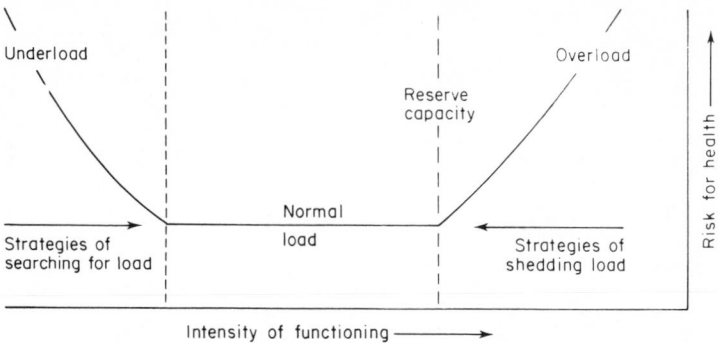

Figure 5.1 Physiological systems can function at variable degrees of intensity. Functioning beyond certain values means overload and underload. Physiological overload means using reserve capacity. Built-in strategies of shedding and searching for load safeguard functioning around physiological setpoints within tolerance limits

part of the capacity could therefore be called emergency capacity. Functioning at increasing levels of intensity implies increasing effort and generally also requires increased motivation. Thus some kind of threat is often needed in order to go on functioning at levels of overload. Such threats might emerge from job insecurity, ambition, or fear of being punished.

Normally, functioning within the limits is maintained by built-in strategies of either shedding or increasing load, and these are involuntarily triggered when a limit is exceeded. In cases of information overload it means that the production of behaviour is less controlled; in the opposite case of underload, one feels bored and looks and/or moves around in search of more information to be processed. Also, spontaneous information generation such as worries, plans, daydreaming, can occur in place of the flux of information imposed by work.

However, if long periods of functioning at higher levels are imposed, the setpoint of the level of functioning of the system might move towards an increased level, an adaptation phenomenon. In that case the limits at which strategies of shedding and of increasing load spontaneously are triggered will also be moved (Figure 5.2). The system is then functioning at a pathological setpoint equivalent to fever in relation to normal body temperature. In the opposite case, as has already been said, a lower level of vigilance will be maintained as a setpoint; this may be induced by isolation, unemployment, retirement, etc. Once a setpoint has been displaced, a shift back to a 'normal' intensity level of functioning is difficult to bring about without training and treatment. A change to working permanently within the region of reserve capacity can also be caused by other mechanisms. Thus, if parents encourage their children overmuch to put extra effort into sport and work, bypassing

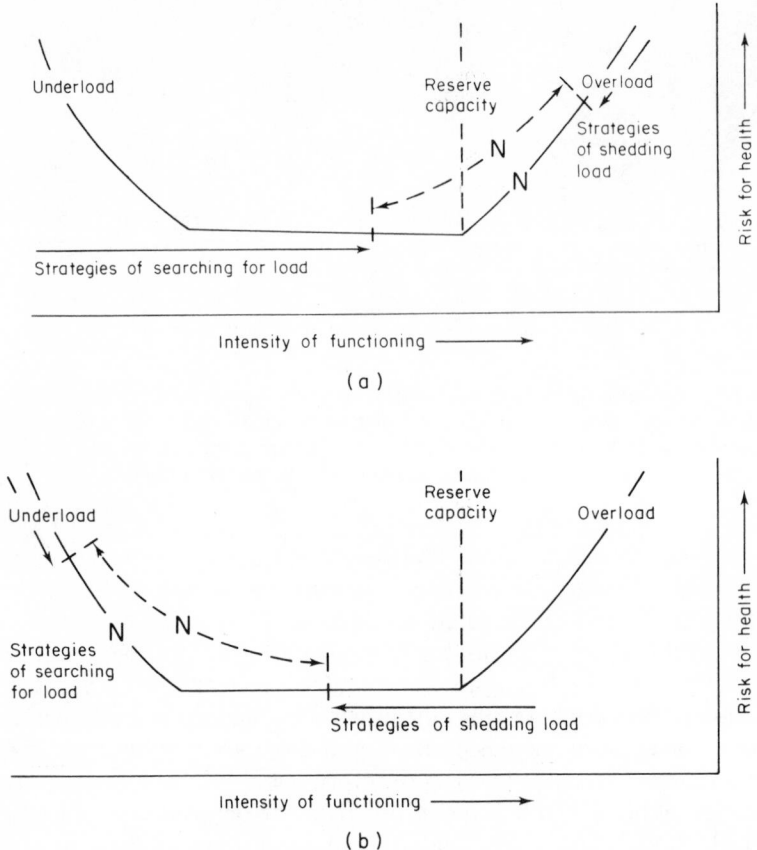

Figure 5.2   Changed setpoints of intensity of functioning. (a) Typical curve of an 'addict' to mental overload. Additional load is searched for even when working in the range of normal intensity of functioning. Reserve capacity is used continuously. (b) Typical curve of an 'addict' to mental underload. Shedding of load occurs even when there is no risk for health. Underloading situations are accepted without a search for additional load

the physiological warning signals can become an aim in itself. One becomes 'addicted' to working within the range of overload.

The concept of intensity of functioning of the brain can be compared with vigilance and arousal levels. The inverted U-shaped curve expressing the correlation between efficiency and arousal is well known. Efficiency is increased to a maximum with increasing arousal. A further increase of arousal will result in a drop in performance. The tolerance lines of Figure 5.1 have been located on the left side of the curve and reserve capacity is used around the top of the curve, as is shown in Figure 5.3. Maximal performance

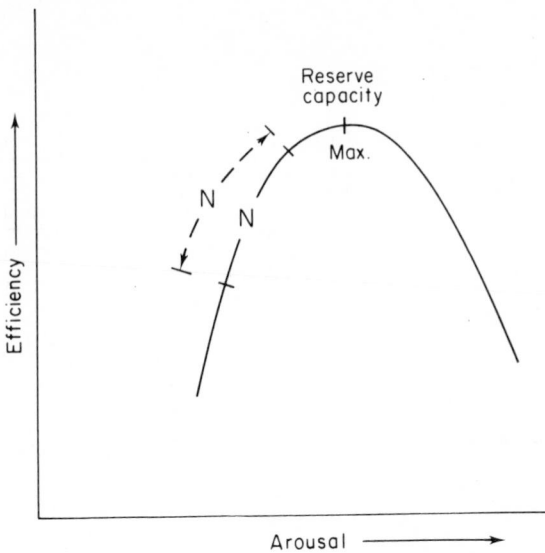

Figure 5.3   Maximal mental output is assumed to be obtained using reserve capacity

and social success are often paid for by a sudden breakdown when reserve capacity has been used to its end.

Mental work capacity

Mental work is a part of the total activity used to control an individual's behaviour production system. Distinction can be made between deliberate (conscious) and involuntary control. Control at the biological level is mainly involuntary. Mental work could be defined as all the deliberate control activities carried out within an individual's behaviour production system. Mental work capacity can then be seen as the capacity available for exercising deliberate control.

Figure 5.4 shows a model of mental work capacity which includes a buffer. A source supplies a buffer, which is linked with a demand. A is the rate of supply and B the rate of demand. If B is greater than A it could be called overload, and if A is greater than B underload. The buffer function belongs to a higher organizational level than the physiological functions. The combined functioning of physiological subsystems, acting as supply systems, represents the 'source'. Physiological systems can function at three recognizable levels: no special effort required, effort required, and emergency state. When B is less than the demand, it leads to an increasing

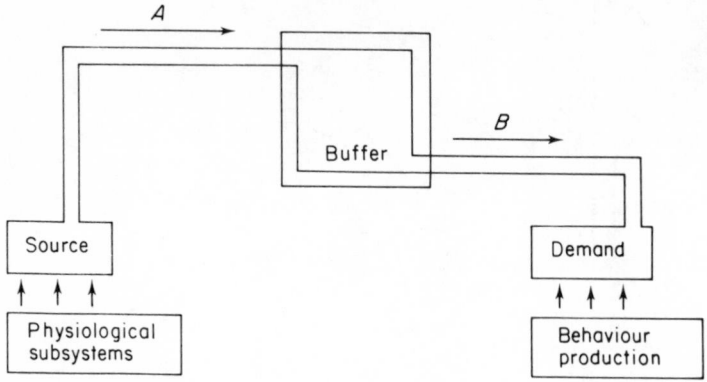

Figure 5.4   Buffer function of mental capacity needed by the behaviour production
processes

lack of deliberate control exerted during behaviour production. When B equals the demand, then all necessary control can be exerted, including deliberate control. Experimental evidence is given elsewhere about what happens when deliberate control necessary to meet the requirements of adapted behaviour is no longer possible (Kalsbeek, 1968). The sequencing of behaviour patterns, which constitutes a behaviour, is no longer based on deliberate choices but is more *passively* provoked by conditions such as momentary inputs, laws of association, mood, etc. This might result in a behaviour which decreasingly meets the requirements of the momentary state of a situation. As a result, the risk of errors, omissions, and blundering will increase, threatening good performance and safety. Perceived differences between actual and required qualities of behaviour as a product may also incite sudden changes in the functioning of physiological subsystems. They also can be taken as a warning signal that it would be better to stop and take some rest.

Effects of fluctuations in mental workload

Because of the buffer function shown in Figure 5.4, a functional overload, i.e. A less than B, becomes a physiological overload (Figure 5.1) only after a long duration. In the opposite case, where B is less than A, there is again no immediate effect on the intensity of functioning of physiological subsystems as long as the capacity of the buffer is not fully employed. Shorter durations of functional overload are often subjectively experienced as stimulating. Shorter durations of functional underload enable the reloading of the buffer to occur. In other words, they serve as a recovery period after short periods of overload (see Figure 5.5). So, hypothetically, there would be no medical problem of mental load if, after each period of functional overload, there followed a period of underload which was long enough to recharge the buffer, even when the physiological systems remained within their normal limits of intensity of functioning. When intermittent periods of underload are not long enough for this, a following period of overload can cause a

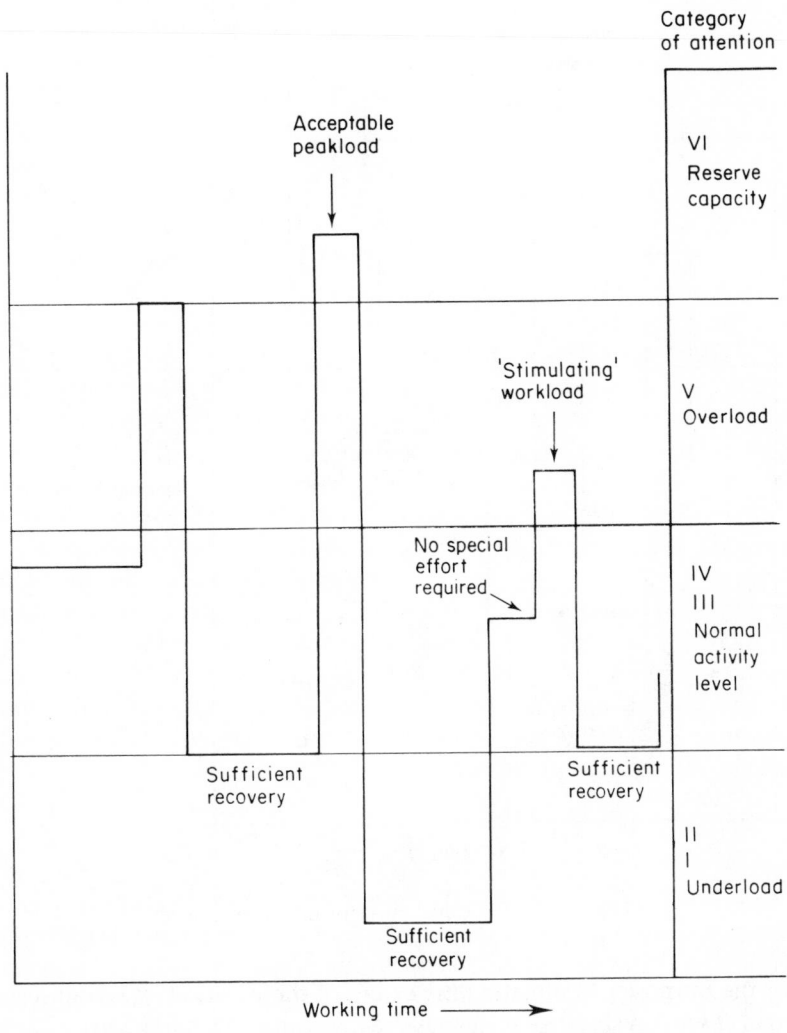

Figure 5.5   Individual ideal curve of mental workload

physiological overload (e.g. nervous breakdown, cardiac attack). This is illustrated diagrammatically in Figure 5.6.

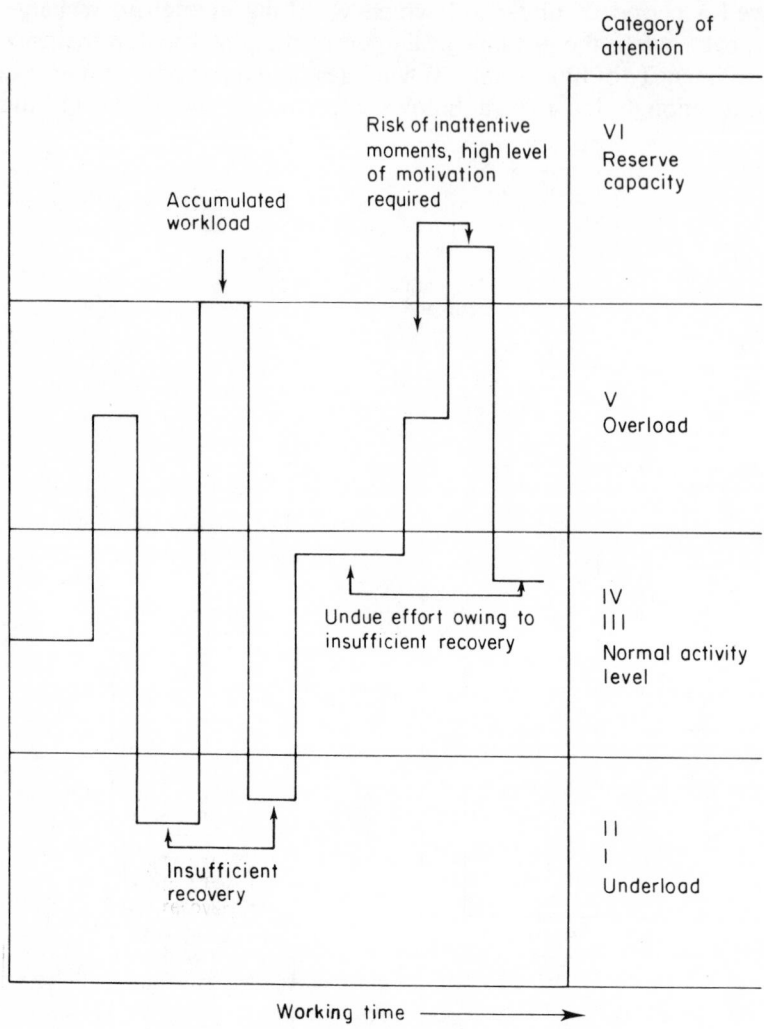

Figure 5.6   Individual curve of mental workload

For the purpose of estimating the extent of these periods when observing the man at work a six point scale was used, as can be seen in Figures 5.5, 5.6, and 5.7. In the upper two categories, corresponding to the maximum attention workload, no moments of conscious control are left for things other

VI Using reserve capacity (danger for health)	Information-handling workload *under pressure*. No moments of attention free to handle information other than from the actual task. *Example*: Peakload in air traffic control. Driving an ambulance in high traffic. High motivation required.
V Overload if long endurance time	High information-handling workload. Tasks asking for continuous control. No moments of attention free to handle information other than from the actual task, but no special motivation required. *Example*: Difficult positioning tasks. Driving in high traffic.
IV Normal level of mental activity	This requires frequent, conscious attention. Moments free to handle information other than from the actual task. *Example*: Contact with other workers, looking around, time for his own personal preoccupations (this could only be done at special moments depending on the performance). *Example*: Putting nuts and bolts together which do not properly fit.
III Normal level of mental activity	Tasks permitting quite frequent attention to information other than the actual task, but at special moments task performance requires complete attention. *Example*: Carefully carrying a glass tube on shoulders.
II I Underloading	Attention is required only incidentally. Tasks comprise subroutines without incidents. Repetitive work without targeting problems. Tasks requiring only superficial attention, routine work.

Figure 5.7   Examples of information workload categories

than the task. The worker's attention is entirely absorbed by the task. In category VI there is an additional factor which could be called 'work pressure'. Even the right man can only perform this task when highly motivated. In category V the worker's attention is again entirely absorbed by his work, but no special motivation is required.

Categories IV and III represent normal attention workload. In category IV, although frequent conscious attention is required by the task, there are moments free to communicate with co-workers, to look around, or to devote attention to preoccupations. These moments cannot freely be chosen but depend on the performance. In category III, the opposite is true; although there are frequent moments of conscious control available to pay attention to

other things, at special moments (depending on the performance), the task
requires complete attention. Categories II and I represent underloading. The
task does not occupy enough moments of conscious control to keep up the
required level of mental activity to maintain the desired level of wakefulness.

Units of mental workload

Mental workload has been defined above as the exercising of deliberate
control on the individual's system for the production of behaviour. The
general procedure in behaviour production has been defined as a sequencing
of behaviour patterns in accordance with sequentially chosen and triggered
control programmes. Control programmes, however, correspond to more
basic behaviour elements such as reaching for an object, or to more elaborate
ones such as driving home and other general routine performances. A
routine programme is able to control the sequencing process of the
behaviour system when sequential uncertainty is low. Only one 'deliberate'
choice is needed to trigger a total complicated programme, the basic elements
being selected by 'routine choices'.

The sequencing within a complicated programme initiates a fixed
programme, comparable to automated dialling by pushing one button
instead of engaging a full set of buttons each corresponding to one character
of a subscriber's telephone number. Only if there is an unexpected event in
the evolution of the situation does the elaborate programme have to be
stopped and another one, which fits the new situation, require to be
triggered.

The distinction between routine and deliberate choices might be expressed
in terms of accessibility of data elements in the brain. For this purpose it is
assumed that data elements differ with regard to their state of readiness for
utilization. Depending on their state, some data elements have a low
threshold and can become operational at lower cost in energy or effort than
others. Selection of data elements is assumed to be brought about by
reducing their threshold values to zero. This operation incurs a cost to the
individual. Roughly four threshold categories can be distinguished:

1. Data elements which are operational whenever a person is awake.
    These are such as concern personal effects arising from age, sex,
    maturity, family composition, day of the week, peace or war, etc.
2. Data elements with a low threshold; a kind of store, selection from
    which can be made by simple choices and be automatically provoked
    by momentary common sensations.
3. Data elements with higher threshold. These data have to be
    deliberately chosen, a relatively costly operation.
4. Data elements which exist but are not available for use by the

behaviour production system. This is because only a limited number of sets of data could be accessed from all the existing data. They are, so to speak, comatose.

The distribution of data elements in these four categories can change from one moment to another. Data elements belonging to category 4 can become available only after a redistribution. Data elements belonging to categories 2 and 3 move into category 4, and *vice versa.*

A redistribution can be brought about in two ways: it can be spontaneously provoked by momentary events, for example by seeing a face, experiencing a string of associations, or by the mood a person is in; and secondly, it can be intentionally brought about by a deliberate choice.

A sequence can be made more flexible by preselecting a subset of data elements (programmes, data, and data structures) corresponding to highly probable changes in a situation, in effect by lowering the thresholds of accessibility. The subsequent selection of a programme from such a stored subset composed of data elements can be 'provoked' by well-defined situational conditions. The contribution to mental workload of such 'routine choices' is assumed to be low. More generally, it is assumed that data elements belonging to an individual's behaviour production system differ with regard to their accessibility. Of the total number of data elements belonging to an individual's store, only a limited part can be operational at a time. The momentary accessibility differs, following a scale from being preselected, as above, to being not accessible. Thus distinctions can be made between low threshold availability, i.e. in the buffer, higher threshold availability, and, even though present, not available for use in behaviour production. The hierarchy of accessibility can change from moment to moment. This change can be brought about by crossing over from one situation (e.g. eating, working, car driving, fishing) in the framework of which behaviours are produced to another. An evident alteration in perceptual input (e.g. change of location, environment or company) can provoke this change. Redistribution of accessibility in this way can take place rapidly and at low mental cost.

The mental load arising from voluntary redistribution depends both on the quantity of data elements which have to be changed in accessibility and on their initial threshold values. Voluntary redistribution of threshold values can constitute a heavy mental load and is time-consuming. This is even more the case when a voluntary distribution has to be brought about in conjunction with an involuntary one, such as a redistribution provoked by a momentary event. Thus the model developed above, instead of assuming different kinds of choices such as deliberate, conscious and routine choices, speaks of threshold values in the accessibility of data elements, these having to be addressed in order to play a role in behaviour production. The

voluntary changing of threshold values becomes an important *operation* with
regard to mental load. It is an operation brought about by the physiological
subsystems. *In this way information loads in behaviour production can be
linked with physiological load.*

Interfaces in man–machine systems

Human behaviour is a function in man–machine systems. It has to meet the
requirements derived from the objectives and constraints of these systems.
Division is made between the artificial component (machine, process, etc.)
and the biological component (man). At the input interface between both
components, signals conveying information about the artificial part are
transformed into physiological signals (perception). Changes in the variables
of the artificial part are rarely directly observed. Commonly, they are
artificially measured and translated into information presentation. How this
artificial component is structured and works is not directly observable.
Nevertheless, incoming information about the state of variables can only
lead to appropriately adapted working behaviour if it is rightly interpreted. It
constitutes what has been called above the mental image. Information about
variables becomes meaningful only if interpreted against the background of a
mental image.

   The acquisition of an adequate mental image is therefore a 'must'. This
can be brought about by, for example, teaching and explanation. Even if
what has to be done is conveyed by very simple instructions (of the kind: if
the lamp is red, you pull that lever), the worker forms his own ideas about
what is behind this information (he makes his own mental image). This leads
to the following conclusions. Firstly, structure and laws of functioning have
to lend themselves to adequate representation in a mental image. Secondly,
the formation of an adequate mental image has to be appropriately guided.
Finally, the structure and pertinence of the presented information has to be
in accordance with the mental image actually present in an individual
worker. However, it was stated above that the data elements, although they
might be present, are not necessarily always operational. Further, it was
stated that elements which have been preselected can, at a particular
moment, be automatically withdrawn easily and rapidly by actual inputs.
Finally, the actual choice of which elements belong in the preselection area
(the buffer) is again mainly conditioned by the perceived situation. This leads
to the conclusion that, for error-free and rapid response, information
presentation has to be such that adequate distribution of data elements to the
preselected store is initiated as well as the adequate selection of elements
from that preselected set. Every decision, however, should not arise from a
buffered set. Reference to Figure 5.1 will show that underload is as bad as
overload. Hence, on occasion, the use of data elements with a higher

threshold level, which cannot simply be provoked but have to be chosen, is necessary. This also means that from time to time a voluntary redistribution of accessibility of data elements has to be made. Nevertheless, one still has to take care that information presentation is such that operations like 'choosing' and 'distributing' do not have to be performed in concurrence with selections initiated by perceptual input, as this is a likely source of errors. In practice, the designer of man–machine systems has to compromise between machine concepts of efficiency and humanly interesting work.

At the output interface, control is exerted on the artificial part of the system. Changes in the artificial part are rarely brought about through direct control. Knobs, handles, and speech constitute interfaces between the artificial and the biological components. Movements brought about by muscles exert a force on knobs, wheels, and handles. Finally, at the interface between chosen programmes and muscles, within the operator, information is transformed into energy.

Thus the voluntary control of information inputs could be regarded as an output, i.e. part of behaviour *as a product*. Even when the sequential uncertainty in the set of microprogrammes is low or absent, their realization by sequential innervation of muscles and corresponding exertion of forces can require a tremendous amount of training before becoming a routine. An example is the playing of music, or other activities which imply manual or other dexterity.

Physiological structures and laws determine which sequences are more natural and will therefore more easily become a routine. Less natural sequences will require voluntary control during a long period of training. Errors are highly likely to occur in cases of stress or panic. Such sequences, therefore, have to be avoided in man–machine systems. A wrong choice of a subsequent programme is called an error. An error occurring within a rightly chosen programme is called a mistake.

Voluntary control during routine performances is more akin to monitoring. The intensity of this monitoring control depends on how 'natural' the sequential innervation of muscles and levels of forces are, and to what extent they are overlearned. Structure, location, etc., in configurations of input and output instruments have to correspond because the perception of input instruments acts directly on the distribution of threshold values of output programmes.

Emotions and behaviour production

A person's mood influences the distribution of threshold values of data elements (programmes, data, data structures). The mechanism by which this change is brought about can be conceived of as if the information–energy transformation at the output interface involved a glandular instead of a muscular activity. Glandular activity, however, does not change the external but rather the internal situation.

In a man–machine combination the distribution of thresholds provoked by emotional changes often interferes with the distribution required by perceptual input and deliberate choices. Moreover, changes of internal emotional states take place gradually, with a large time constant. Conflicts are likely to arise, therefore, which might lead to unsafe conduct. It seems that emotions in a natural environment prepare an individual for 'explosions' of energy, attack, flight, sex, aggression, etc. If these 'explosions' do not follow, internal metabolic problems may occur. On the other hand, peak loads of information can initiate momentary states of irritation and aggressiveness. Often these stop as soon as the flow of information returns to normality.

If a person regrets his loss of temper, this can be seen as an emotional after-effect of longer duration which again acts upon the distribution of thresholds of data elements. A biased selection of data elements which is either accessible or not results for a certain time in a biased, i.e. unrealistic, conception of the world in which the person lives, as is the case, for example, during depressive or paranoid states.

Finally, a sequential effect in information and emotional peakloads needs to be mentioned. After an emotional peakload, for example the death of a beloved person, a dense flow of information can be experienced as alleviating. In the opposite sequence, however, an emotional peakload can imply a great risk for health if it comes after a period of intensive information load, i.e. when reserve capacity becomes exhausted. Emotions can induce reverberating information flows which occupy information-handling capacity. It is an interesting question whether the curve of Figure 5.1 would also apply to emotional loads.

Mental workload and microtiming

The sequentially chosen and realized microprogrammes have also to fit the temporal constraints defined by the man–machine system. This implies planning with regard to the time dimension.

Perceived differences between subjectively predicted and objectively imposed (micro) timing can lead to less accuracy, agitate the physiological subsystems to function at higher intensity, and/or create a general state of nervousness and tension.

Diminished accuracy increases the risk of unsafe conduct and can initiate feelings of failing. Physiological functioning at higher intensity for longer duration means the exhausting of reserve capacity. Feelings of failure, and of nervousness and tension for longer periods, are important risk factors to health and well-being (especially in relation to arterial and coronary diseases).

Actual, but not perceived differences between subjectively predicted and objectively imposed time schedules do not constitute a workload during the behaviour-planning phase.

Reference

Kalsbeek, J. W. H. (1968). 'Measurement of mental work load and of acceptable load: possible applications in industry.' *The International Journal of Production Research*, **7**, No. 1.

Stress, Work Design, and Productivity
Edited by E. N. Corlett and J. Richardson
© 1981 John Wiley & Sons Ltd

Chapter 6

Physiological Measures of Mental Load

Helmut Strasser
Technical University
Munich, Germany

Enormous efforts have been undertaken in order to ascertain to what extent physiological parameters may be used for the assessment of work load. In this context, reference must be made to the symposium in London in 1971 on 'Heart Rate Variability and the Measurement of Mental Load' and the meeting in Darmstadt, also in 1971, on 'Assessment of Work Load in Air Traffic Control Tasks' (see Rohmert, 1971). In the same year a colloquium on 'Heart Rate Recordings and their Application in Work Study' (see e.g. Ehrenstein, 1973; Rohmert, 1973) was held in the Max-Planck-Institute for Agriculture in Bad Kreuznach (GFR). The aim of this meeting also was to demonstrate the limits of applicability of heart rate measurements and the possibilities of interpreting the results either alone or in combination with other physiological criteria in ergonomic research.

It is almost impossible here to give a brief but relevant excerpt from all the knowledge. However, a short report of the state of the art from the author's point of view will be presented, followed by some ideas which may bring us a small step further in our still scanty knowledge of the methods of assessing non-physical work load which can be applied in field studies.

The stress–strain relationship, in which the individual's capacities and abilities are included (see, amongst others, Rohmert, 1973), demonstrates the necessity to measure physiological variables when trying to assess stress and strain. But there still remain problems in interpreting the results which cannot essentially be tackled without psychological or sociological tools, as other chapters in this book have shown.

Insofar as continuous physiological measurements are concerned, heart frequency is mainly used as the integral physiological criterion for dynamic and static work, for heat, and psychological or emotional stress. But in order to distinguish between the sources of heart rate increase it is also necessary to register other parameters. For instance, it is possible to produce a qualitative analysis of the stress involved by means of oxygen intake, metabolism of catecholamines, or physical measures like temperature or static forces. From experience we know that heart rate acts somewhat like an integral over several factors which determine physiological homeostasis. As an indicator of the *milieu interne*, it can be registered in a relatively simple way. It therefore seemed advantageous to also use heart rate measurements when trying to assess factors like mental load and fatigue. But heart rate is not an especially suitable parameter for the quantification of stress or mental load in monotonous working conditions which do not require any noticeable motor activities from the subject. Values measured during resting and during working phases do not differ systematically (see Strasser, 1974a; Mulder and Mulder-Hajonides van der Meulen, 1972). Because metabolism is not essentially increased in situations which demand concentration and mental load, heart rate does not react. In addition, changes of heart rate resulting from moments of psychological stress are mostly of very short duration. They cannot therefore be detected when heart rate is averaged over one minute or even longer time intervals.

But what is to be done if heart rate does not change at all under mental load? It is necessary to apply measurement techniques which are not so rough and which can indicate dynamic changes in heart rate control. Such indications may be found in the micropicture of pulse rate, that is to say in the cardiotachogram. Therefore in order to assess load on the human being in mainly mental tasks and non-physical load situations, manifold different scoring methods of the so-called 'sinus arrhythmia', (Kalsbeek, 1967, 1971) have become increasingly important.

Some years ago, Kalsbeek (1973)—who may be called one of the fathers of sinus arrhythmia—put the question to himself, 'Do you believe in Sinus Arrhythmia?'. Of course, this question was not simply the result of a possible eventual scepticism concerning the parameter sinus arrhythmia, but derived also from discussions with practical men. Perhaps in the first enthusiasm they thought that through sinus arrhythmia they had been given a new general, valid and powerful measurement for assessing mental strain. Yet they might have overestimated the predictability of this parameter for the individual's load. If someone is occupied intensively with this question he may come to the conclusion that sinus arrhythmia as a scoring method should be valid for rather undefined and diffuse situations, that is to say for the quantification of a global mental load. In real work situations we seldom meet exactly defined mental load tasks. Additionally disturbing effects of ambient temperature changes, different body postures (Hanson and Jones, 1970) or extra energetic work are mostly involved. More than that, the relevant information flow, i.e. the load from the job which is to be quantified as the main cause of the individual's strain, is hidden by a more or less high load of situational information. Kalsbeek (1973) therefore came to the conclusion that sinus arrhythmia at least is 'an indicator of the proportional occupation of an individual's single-channel capacity during rest and work'.

Some examples of the behaviour of heart rate under physical and non-physical stress are shown in Figures 6.1 and 6.2. In Figure 6.1 two recordings are presented which show that increased heart rate and simultaneously decreased variability are the results of increasing energetic work. Figure 6.2, besides other registrations, shows beat-to-beat heart rate during rest and tracking as a non-physical work load. Arrhythmia decreases during work while heart rate even shows a slightly lower level during the load of the tracking task. Brief variations of mental load during a task can be shown by a scoring method for arrhythmia which is derived from beat-to-beat heart rate, or the R to R intervals in the ECG. Here the sum of absolute differences between succeeding R–top–intervals was used. In Figure 6.3 a correlation between tracking error—in the upper trace—and the scores of sinus arrhythmia integrated over one minute can be seen. Sinus arrhythmia seems to indicate the effort of the subject which he brings into action in order to fulfill the demands. When the subject tracks well, integrated absolute error gradient is low and sinus arrhythmia is suppressed too. When tracking proficiency is poor, sinus arrhythmia is high.

But now to some statistically relevant results, found in laboratory studies during prolonged sessions. The upper portion of Figure 6.4 shows heart rate profiles and the lower portion corresponding sinus arrhythmia values during tracking sessions of about two hours. The measurements of the second hour were identical in respect to the work–rest schedule. Prolonged working caused decreasing heart rate and slight elevations of sinus arrhythmia, scored as shown before. Increased demands on motor components in two tracking tasks are most visible in an elevated heart rate. To a certain degree heart rate level determines sinus arrhythmia. An elevated heart rate is accompanied by more suppressed irregularities and a lowered heart rate brings about increasing arrhythmia. Figure 6.4 also shows that, with no obvious differences in heart rate, sinus arrhythmia during rest and all working phases is more depressed in the group of older subjects than in the young subjects.

Other tests with groups of different ages (Strasser et al., 1973) and experiments with and without a central nervous stimulant (Strasser and Müller-Limmroth, 1973b,c) demonstrated that no complete negative correlation between heart rate and irregularity of cardiac activity exists. Sinus arrhythmia was nearly constant at the same level for several hours tested, whereas heart rate fell rapidly. The results indicated that older subjects could take almost the same level of stress as young, physically fit subjects when supported by the intake of central nervous stimulating drugs.

When measuring heart rate and sinus arrhythmia profiles in laboratory studies a more or less marked decline of heart rate with increasing test duration was always found. But this tendency to decline, as well as the level of sinus arrhythmia, could be modulated by different factors. For instance,

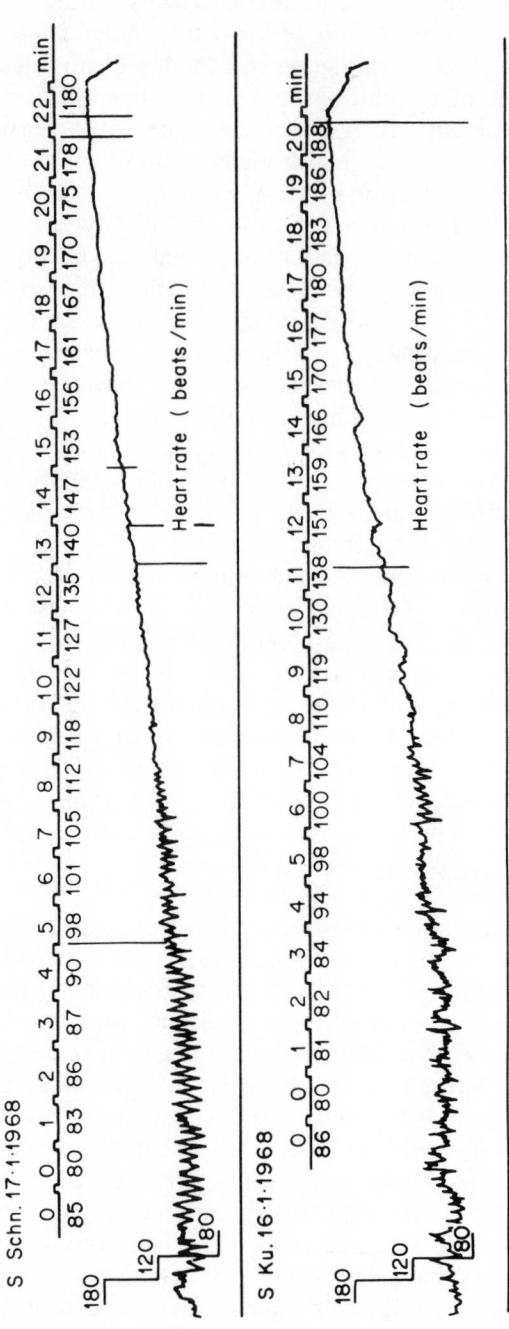

Figure 6.1 Beat-to-beat heart rate of two subjects under the influence of a continuously increasing physical load by means of a bicycle ergometer (After Ehrenstein, 1973)

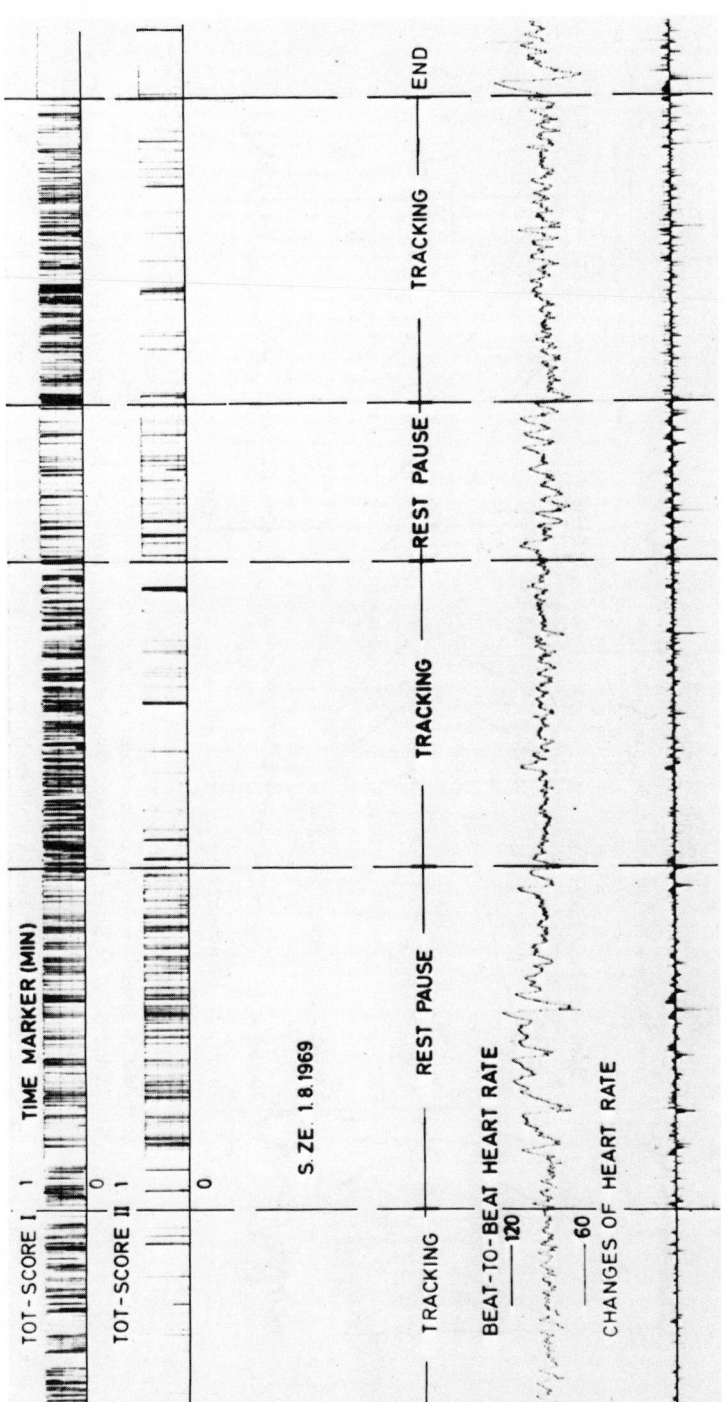

Figure 6.2 Registration of a cardiotachogram (beat-to-beat heart rate) in rest (rest pauses and end of test) and during a tracking task as a non-physical work load. Besides the cardiotachogram, the changes of heart rate from each single beat to the next (changes of heart rate) and different Time-Off-Target scores are shown. By means of the different TOT scores performance was measured (After Strasser, 1974b)

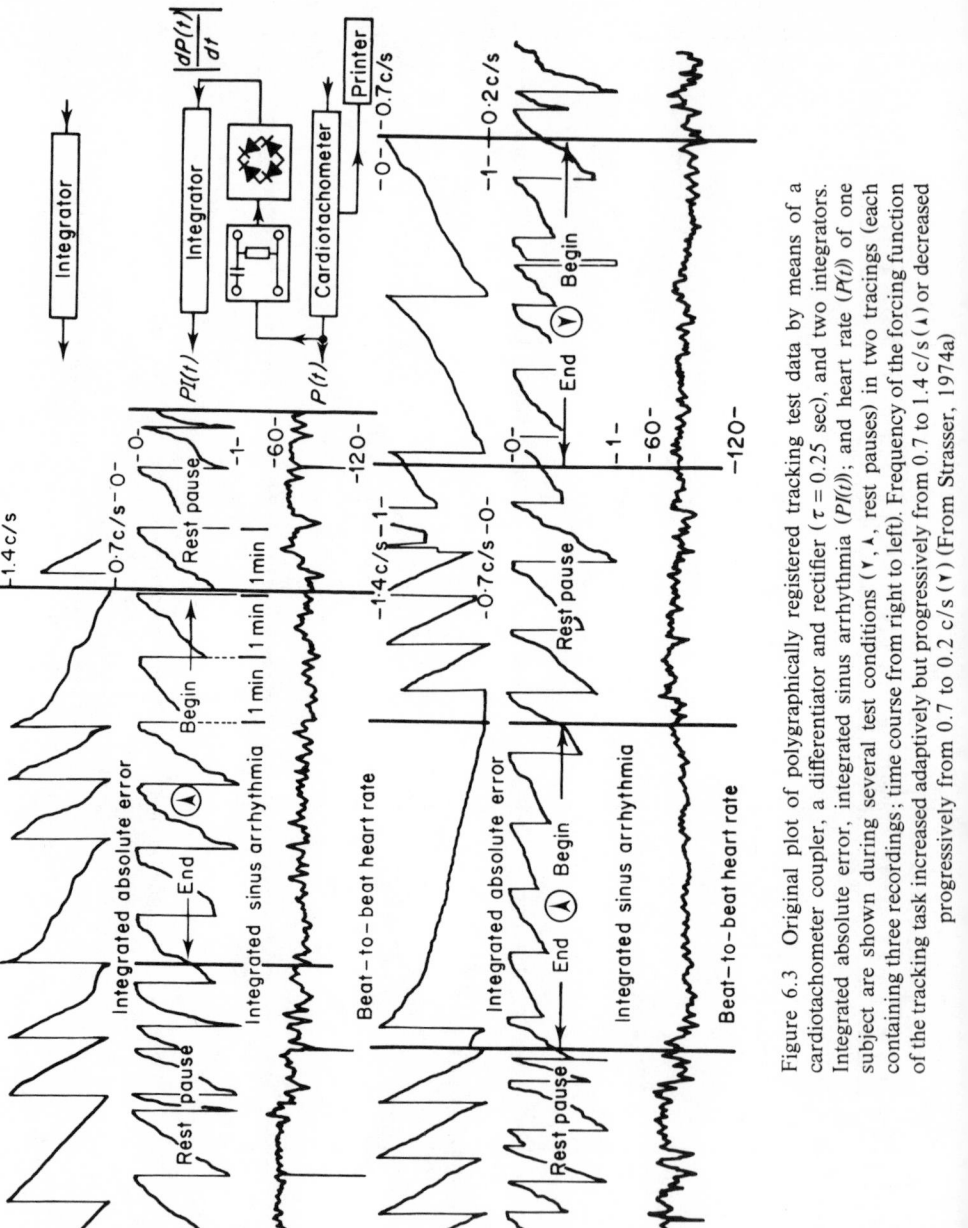

Figure 6.3 Original plot of polygraphically registered tracking test data by means of a cardiotachometer coupler, a differentiator and rectifier ($\tau = 0.25$ sec), and two integrators. Integrated absolute error, integrated sinus arrhythmia ($PI(t)$); and heart rate ($P(t)$) of one subject are shown during several test conditions ($\blacktriangledown$, $\blacktriangle$, rest pauses) in two tracings (each containing three recordings; time course from right to left). Frequency of the forcing function of the tracking task increased adaptively but progressively from 0.7 to 1.4 c/s ($\blacktriangle$) or decreased progressively from 0.7 to 0.2 c/s ($\blacktriangledown$) (From Strasser, 1974a)

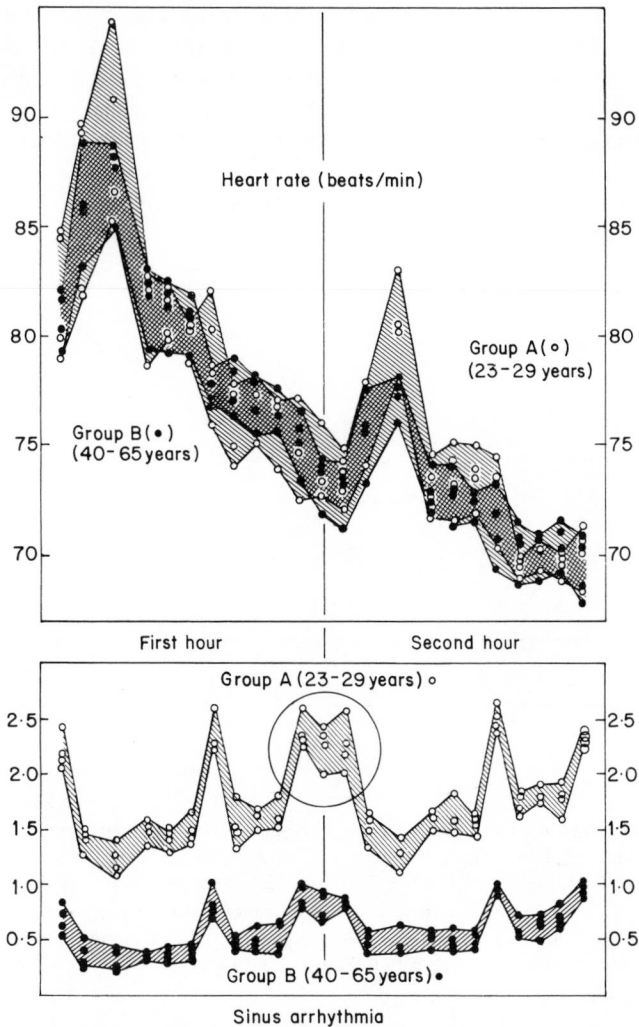

Figure 6.4   Profiles of heart rate (upper panel) and sinus arrhythmia (lower panel) of two test groups. Test group A consisted of 15 male subjects (students) aged from 23 to 29 and group B of 10 male subjects (employees) aged 40 to 65. Each group was tested four times under the same conditions for two hours each. Solid lines represent the extreme means of each group. The test programme consisted of four different tracking tasks and rest pauses, which were repeated each day (first and second hour). For details see Strasser *et al.* (1974)

psychopharmaca when used for stimulating reduced the decline, but when used for tranquillizing (Strasser and Müller-Limmroth, 1973a) enhanced it. Such profiles can well be interpreted in accordance with the hypothesis of

Rutenfranz, *et al.* (1971). According to this hypothesis (illustrated in Figure 6.5), heart rate can be regarded as being composed of several factors. Fractions called 'static and dynamic parts', 'intentional basic tension', 'basic resting level', and 'psychogenic factors' can be distinguished. When applying this hypothesis to his results the author usually says that the fraction 'intentional basic tension'—which in Figure 6.5 is represented by the vertical striped area—seemed to diminish with the length of the sessions, thereby resulting in an overall decrease in heart rate.

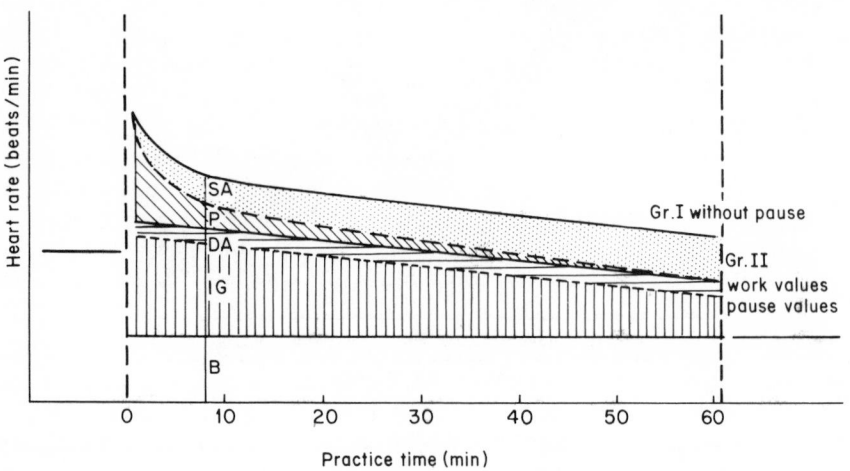

Figure 6.5   Schematic diagram of the changes of pulse frequency 'fractions' on practising a tracking task, after different kinds of practice. B, basic pulse frequency at repose while sitting (working position) after the trial. IG, increase of pulse frequency as the result of the intentional basic tension. DA, increase of pulse frequency as the result of dynamic work. P, increase of pulse frequency as the result of psychogenic, mostly emotional reactions. SA, increase of pulse frequency as the result of static muscle work (From Rutenfranz *et al.*, 1971)

We think it necessary to obtain similar working profiles as physiological representations of workplaces under different conditions, for instance for normal and 'enriched' or 'enlarged' jobs, for places with and without the disturbing effects of noise, etc. However, such profiles should be established by registering heart rate and sinus arrhythmia not from one subject after another, but simultaneously from a group of workers on the same job. By this procedure the disturbing effect of fluctuating situational information flow at different times will be reduced.

Of course, other physiological variables which can indicate mental stress reactions (Berkhout, 1970; Laurig and Rohmert, 1974), as for instance EMG (electromyogram) or GSR (galvanic skin response), should also be applied in field studies. But as far as the author knows, it has not yet been demonstrated whether they are not too complex to interpret. Other variables which have been proved in laboratory studies cannot be used in field studies because of technical difficulties. For instance, the EEG is a very nice parameter to show slight changes in the activity of the central nervous system. Also by means of evoked potentials, that is to say stimulus-related specific changes in the EEG, it is possible to demonstrate different stages of vigilance (Fruhstorfer and Bergström, 1969). In this respect Klinger and Strasser (1972) and Klinger (1973) have shown clear effects of alcohol or of relevant and irrelevant information feedback on proficiency and mental load in tracking tests. There has been much research in the field of evoked potentials (Finkenzeller, 1969; Keidel, 1973; Blom, 1974; etc.), yet this very delicate parameter is not a practical one for use in field studies. The workers on their jobs would be disturbed too much when evoked responses were elicited. Though flicker fusion frequency (see bibliography from Ginsburg, 1970; Schmidtke, 1965, 1973) is a relatively good indicator of alterations of the vegetative and central nervous status, it also can hardly be used while on a job. Determination of the level of catecholamines in the urine (Klimmer, *et al.* 1972, 1973) may be a help, for instance, in order to decide whether an increased heart rate is due to physical or emotional stress (Ulmer, 1973), but we must recall that these measurements are not continuous ones. However, we can hope that we will soon obtain more information on the usefulness of this parameter from Rutenfranz's working group.

With respect to central nervous and vegetative parameters, perhaps it would be valuable to control the physiological state of stressed subjects not only during the job but during periods of recovery, for instance in sleep, by means of around-the-clock studies.

As far as the cardiovascular parameters of heart rate and sinus arrhythmia are concerned, the following points should be stressed. We can continue to measure heart rate and irregularity scores as well-accepted indicators of mental load, but still there remains the problem of interpreting the results. We have to deal with the question of what changes in physiological parameters really do mean. Is it already dangerous to have heart accelerations of about 10 beats per minute during a work phase or not? Does this increasing or decreasing heart rate indicate a special stress which will result in decrements of performance? Are there correlations between physiological parameters and performance scores? Must there be a correlation or not? Because there need not be a correlation in every case, it is urgently necessary to register physiological parameters too. This may be illustrated by means of results obtained during one of the author's series of laboratory sessions (Strasser and Müller-Limmroth, 1974). Hypoxia resulted in a very high increase in heart rate, but no decrements in tracking performance, down to a grade of hypoxia stimulated by about 12% oxygen in the inspired air (equal to a height of about 4200 m), could be found. This result could be very interesting in respect of stress research. No doubt there was a very strong stress in the tests, but no reaction as evidenced by an operational change. In industrial processes there can be situations where the striking features are not the decrements in performance output of a man–machine system but stressed and angry workers. It goes without saying that it would be interesting to research such working situations. Nowadays it is no problem to measure heart rate and to calculate sinus arrhythmia scores by tape recorders during the whole shift time of work and to analyse the recordings with a computer.

Another question that arises in this respect is that of interpreting the results. Concerning this problem it should be said that physiological reactions in field studies must be calibrated by laboratory experiments. We are not in the fortunate situation of being able to say that this or that degree of change is dangerous or will be unhealthy. Vogt *et al.* (1973) have shown how to carry out a calibration. In Figure 6.6, partitioning of observed heart rates into motor and thermal components is demonstrated, and by means of standard laboratory tests, motor and thermal cardiac reactivity indices were obtained. When analysing heart rate measured during actual working

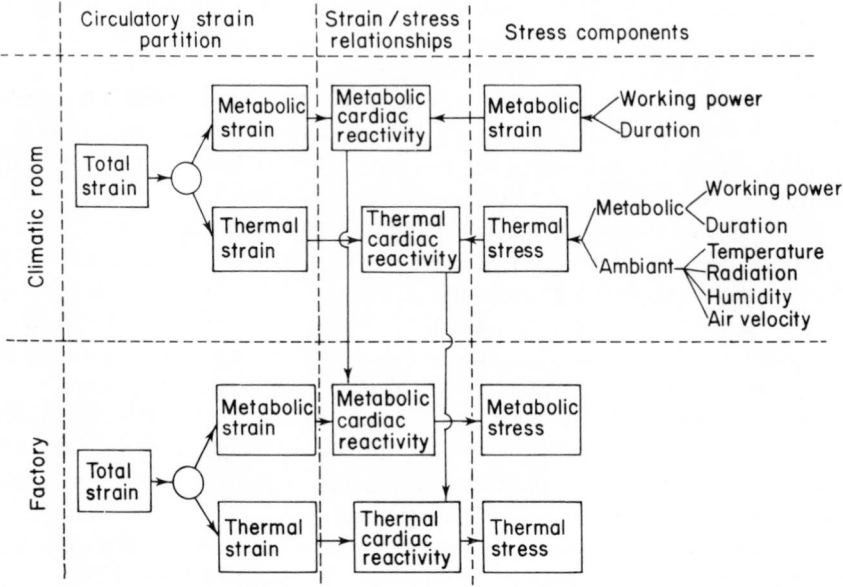

Figure 6.6   Flow diagram of the practical procedure of indirect evaluation of muscular work and environmental heat stresses from the continuous record of heart rate (From Vogt *et al.*, 1973)

situations, total reactions can be split into their components by means of the reactivity indices. What is shown here must not be restricted to physical loads but should also be applied to mental loading tasks.

Another possibility of obtaining information on the meaning of changes in physiological parameters may be provided by the use of adaptive laboratory tests. Research test designs usually use fixed levels of difficulty and record performance and physiological changes, stressing the changes in performance in the results. Adaptive tests seem likely to be more appropriate for assessing mental load: such a design would adapt the test conditions to maintain the physiological parameters constant, or within limits, and investigate the changes in load and the resulting changes in the performance measures.

References

Berkhout, J. (1970). Psychophysiological stress: Environmental factors leading to degraded performance. In *Systems Psychology* (Ed. DeGreene, K. B.). McGraw-Hill, New York, pp. 407–50.

Blom, J. L. (1974). Influence of a nonvisual binary choice task on the visual evoked response. Report of the Laboratorium voor Ergomische Psychologie van de Gezondheidsorganisatie, TNO, Amsterdam.

Ehrenstein, W. (1973). Interpretation von Herzfrequenzmessungen mit Hilfe bekannter Beziehungen zu anderen Kreislaufgrößen. In *Pulsfrequenz und Arbeitsuntersuchungen. Schriftenreihe 'Arbeitswissenschaft und Praxis'*. Beuth-Vertrieb, Berlin, Cologne, Frankfurt, pp. 66–73.

Finkenzeller, P. (1969). Die Mittelung von Reaktionspotentialen. *Kybernetik*, **6**, 22–44.

Fruhstorfer, H., and Bergström R. M., (1969). Human vigilance and auditory evoked responses. *Electroenceph. clin. Neurophysiol.*, **27**, 346–55.

Ginsburg, N. (1970). Flicker fusion bibliography, 1953–1968. *Perceptual and Motor Skills*, **30**, 427–82.

Hanson, J. A., and Jones, F. P. (1970). Heart rate and small postural changes. *Ergonomics*, **13**(4), 483–7.

Kalsbeek, J. W. H. (1967). *Mentale Belastung. Theoretische en experimentele exploratories ter outwikkeling van meet-methoden*. Van Gorcum, Assen.

Kalsbeek, J. W. H. (1971). Sinus arrhythmia and the dual task method in measuring mental load. In *Measurement of Man at Work*. (Ed. Singleton, W. T., Fox, J. G., and Whitfield, D.). Taylor and Francis, London, pp. 101–13.

Kalsbeek, J. W. H. (1973). Do you believe in sinus arrhythmia? *Ergonomics*, **16**(1), 99–104.

Keidel, W. D. (1973). Informationsverarbeitung. In *Kurzgefaßtes* Lehrbuch der Physiologie. (Ed. Keidel, W. D.). Georg Thieme, Stuttgart, pp. 385–401.

Klimmer, F., Aulmann, H. M., and Rutenfranz J., (1972). Katecholaminausscheidung im Urin bei emotional und mental belastenden Tätigkeiten im Flugverkehrskontrolldienst. *Int. Arch. Arbeitsmed.*, **30**, 65–80.

Klimmer, F., Aulmann, H. M., and Rutenfranz J., (1973). Katecholaminausscheidung bei mental belastenden Tätigkeiten im Flugverkehrskontrolldienst. In

Problematik von Arbeitsplätzen mit mentaler Belastung. Pathogene Stäube mit ihren Auswirkungen auf den Menschen. (Ed. Wenzel, H. G. and Tentrup, F. J.). Bericht über die 12. Jahrestagung der Deutschen Gesellschaft für Arbeitsmedizin e.V., Dortmund, 25.–28.10.1972. Gentner, Stuttgart, pp. 69–75.

Klinger, K.-P. (1973). Amplitudenvariationen akustisch evozierter Potentiale in Abhängigkeit von der Signalinformation und belastungsbedingter Ermüdung. *Int. Z. angew. Physiol.*, **31**, 269–78.

Klinger, K.-P., and Strasser, H. (1972). Variations of physiological parameters during defined mental load and rest. *Pflügers Archiv.*, **332** (Suppl.), R82.

Laurig, W., and Rohmert W., (1974). Ergonomische Methoden zur Beurteilung des Teilsystems 'Mensch' in Arbeitssystemen. In *Ergonomie* 2 (Ed. Schmidtke H.). Carl Hanser, Munich, pp. 113–45.

Mulder, G., and Mulder-Hajonides van der Meulen W. R. E. H., (1972). Heart rate variability in a binary choice reaction task: An evaluation of some scoring methods. *Acta psychologica*, **36**, 239–51.

Rohmert, W. (1971). Introduction to: An International Symposium on Objective Assessment of Work Load in Air Traffic Control Tasks. Held at the Institute of Arbeitswissenschaft, The University of Technology, Darmstadt, GFR. *Ergonomics*, **14**(5), 545–7.

Rohmert, W. (1973). Pulsfrequenz und Dauerleistungsgrenze. In *Pulsfrequenz und Arbeitsuntersuchungen. Schriftenreihe 'Arbeitswissenschaft und Praxis'.* Beuth-Vertrieb, Berlin, Cologne, Frankfurt, pp. 21–33.

Rutenfranz, J., Rohmert, W., and Iskander, A. (1971). Über das Verhalten der Pulsfrequenz während des Erlernens sensomotorischer Fertigkeiten unter besonderer Berücksichtigung der Pausenwirkung. *Int. Z. angew. Physiol.*, **29**, 101–18.

Schmidtke, H. (1965). *Die Ermüdung.* Hans Huber, Berne, Stuttgart.

Schmidtke, H. (1973). Mentale Beanspruchung. In *Ergonomie* 1 (Ed. Schmidtke, H.) Carl Hanser, Munich, pp. 256–79.

Strasser, H. (1974a). Beurteilung ergonomischer Fragestellungen mit Herzfrequenz und Sinusarhythmie (Indicatoren von mentaler Beanspruchung und Ermüdung). *Int. Arch. Arbeitsmed.*, **32**, 261–87.

Strasser, H. (1974b). Technisch-physiologische Aspekte der Beziehung Stress–Strain. Eine modell-theoretische Betrachtung. *Arbeitsmedizin-Sozialmedizin-Präventivmedizin*, **9**(10), 212–7.

Strasser, H., Brilling, G., Klinger, K.-P., and Müller-Limmroth, W. (1973). The physiological and operational state of a group of aeroplane pilots under the conditions of stressing tracking tests. *Aerospace Med.*, **44**(9), 1040–7.

Strasser, H., Müller, K.-W., and Adler, A., (1974). Some factors influencing heart rate and sinus arrhythmia in rest and during tracking tasks. *Pflügers Archiv.*, **347** (Suppl.), R24.

Strasser, H., and Müller-Limmroth, W. (1973a). Komplexe Auswirkungen der Faktoren Lärm, Tranquilizer, erschwerte Arbeitsbedingung und Versuchszeit auf eine Pursuit Trackingleistung und das kontinuierliche Puls-zu-Pulsverhalten. *Int. Arch. Arbeitsmed.*, **31**, 81–103.

Strasser, H. and Müller-Limmroth, W. (1973b). Physiologische Veränderungen und Regelleistungsverhalten älterer Probanden während kontinuierlicher Trackingtätigkeiten nach Zufuhr einer zentral aktivierenden Substanz. *Arzneim.-Forsch. (Drug-Res.)*, **23**(3), 406–15.

Strasser, H., and Müller-Limmroth, W. (1973c). Experiments with heart rate and sinus arrhythmia as indicators of fatigue and mental load. *Pflügers Archiv.*, **339** (Suppl.), R27.

Strasser, H., and Müller-Limmroth, W. (1974). Tracking performance and physiological parameters in hypoxia. XXII Int. Congress of Aviation and Space Medicine, Beirut/Liban, 6.-12.10. Abstract of papers, p. 12.

Ulmer, H. V. (1973). Physiologische Grundlagen zur Beurteilung der Arbeitsbeanspruchung mit Hilfe von Pulsfrequenzmessungen. In *Pulsfrequenz und Arbeitsuntersuchungen Schriftenreihe 'Arbeitswissenschaft und Praxis'.* Beuth Vertrieb, Berlin, Cologne, Frankfurt, pp. 41–50.

Vogt, J. J., Meyer-Schwertz, M. Th., Metz, B., and Foehr, R. (1973). Motor, thermal and sensory factors in heart rate variation: A methodology for indirect estimation of intermittent muscular work and environmental heat loads. *Ergonomics*, **16**(1), 45–60.

Stress, Work Design, and Productivity
Edited by E. N. Corlett and J. Richardson
© 1981 John Wiley & Sons Ltd

Chapter 7

A Transactional Approach to Occupational Stress

Tom Cox
University of Nottingham, UK
and
Colin Mackay
EMAS, London, UK

Overview

Lumsden (1975) believes that the concept of stress 'is one of the most significant and integrative concepts ever developed in the social and biomedical sciences, and that its potential as a prime intellectual tool for not only understanding, but also explaining, individual and collective human behaviour and disorders has not yet been fully realised'. However, one of the major problems is that the concept of 'stress' is now widely used in a variety of very different disciplines, such as medicine, psychology, sociology, engineering, and anthropology. Such usage has led to a multiplicity of definitions, statements, theories, and models. These range from mere descriptive and general assertions, where the reader is largely left to guess at the author's meaning, through to complex and sophisticated theories which attempt to account in detail for specific data. Many are *post hoc* and few are predictive.

Three major questions need to be answered. First, is the concept of stress a necessary or useful one; second, in what way can the existing models and formulations of stress be categorized, and are they adequate; and third, if the term is not to be abandoned in what way can it be developed and used? In the remaining sections of this chapter each of these questions is raised and discussed. The first section contains a brief review of the nature of occupational stress, commenting upon some of the factors in the work situation known to be responsible for its occurrence. It suggests that the concept of stress is useful as an 'economic' way of bringing together the literature on the many different problems and concepts related to work. The next section categorizes existing models of stress and evaluates the contribution each type has made to our understanding of the phenomenon. It suggests that certain types are fundamentally inadequate. This section concludes with the elaboration of the transactional model of stress developed by the present authors. Next, the transactional approach is used to investigate various techniques for the management and reduction of stress. Finally, a brief review of research based on this approach is presented, followed by a summary. It is suggested that the transactional and related models could represent a shift towards a new view of occupational problems, in terms of what have been called 'discrepancy' theories.

Occupational stress: its scope and nature

Since the late 1960s an intense debate has developed around the structure of industrial work and the various problems associated with it (see Swedish Employers' Confederation, 1975; Engstrom et al., 1971; Caplan et al., 1975). Effects such as poor job performance (Buzzard, 1973), wasted leisure time (Gardell, 1973; Wilensky, 1960; Frankenhaeuser, 1975), low job satisfaction (Van Harrison, 1976), and increased morbidity and mortality (Carlestam et al., 1973) have all been shown to be associated with bad job design and other related problems at work. In the same vein, it is recognized that an increasing number of the conditions faced by the medical and social services appear to be caused, or exacerbated by problems experienced at work. These are therefore increasingly seen as a threat not only to the quality of working life but to the quality of life in general.

However, some of the statistical evidence concerning the effects of work problems presents a confusing and paradoxical picture. A number of surveys both in Britain and in Europe indicate that the vast majority of people report that they are satisfied with their job (Taylor, 1974). Somewhat by contrast, a report by the Department of Health and Social Security in 1969 shows that illnesses possibly attributable to problems at work, such as 'nerves', debility, and headaches, as well as heart disease and neuroses, are on the increase, while the prevalence of more traditional physical illnesses is decreasing. These data and the recently published 'occupational mortality statistics' (OPCS in 1978) have been commented on in more detail by the authors (Cox, 1978; Cox and Mackay, 1979). The apparent paradox is largely explained away by the inadequate nature of most general survey questions on job satisfaction.

Another relatively recent survey (OPCS in 1976) shows a strong positive relationship between level of reported job dissatisfaction and repeated absence from work. The overall trend indicates that those persons feeling 'rather or very dissatisfied' with their jobs report twice as much absence from work as those 'very or fairly satisfied'. There are at least three possible explanations for these data. First, individuals who are naturally pessimistic and generally dissatisfied may also be lax and not fulfil obligations, including those at work. A second explanation may be that, because of say illness or injury causing absence from work, the individual has a reduced capacity (physical or mental). Work may then become much harder, perhaps resulting in decreased job satisfaction. The third explanation is that high levels of job dissatisfaction may trigger attempts to cope with its causes; one often cited method is to stay away from work. In this context it is interesting to note that the 1972 General Household Survey showed that the highest levels of absence from work occurred in those jobs requiring the lowest levels of skill, where *underutilization of skill* may be a problem.

A detailed discussion of the various job demands that might cause dissatisfaction and absence will not be presented here; they have been adequately reviewed elsewhere (see, for example, Cox, 1978; McGrath, 1970, 1976; Hackman, 1970; Cooper and Marshall, 1976; Buzzard, 1973; Caplan *et al.*, 1975; Mackay and Cox, 1975; Warr and Wall, 1975). However, one study will be mentioned in order to illustrate the wide range of demands that have been implicated. An early study carried out during the later stages of the Second World War (Fraser, 1947) investigated the factors thought to predispose workers to neurotic illness, since this appeared to be accounting for a considerable amount of absence from work. Over 3000 workers, both male and female, were selected for interview and test on a random basis from a much larger group working in the engineering industry. What emerged was a list of factors and circumstances associated with an increased incidence of neurosis. These factors included long working hours, inadequate diet, reduced leisure time, excessive responsibilities at home, inappropriate level of skills for the job, work requiring constant attention but allowing little responsibility, initiative or variety, and environmental factors, including poor lighting. It is probable that many of these factors are interrelated, such as long working hours, inadequate diet, and reduced leisure time. Nevertheless, the list does emphasize the wide range of job factors associated with increased illness and absence. Although this survey was carried out over thirty years ago, and during wartime, much more recent surveys present a depressingly similar picture (Theorell, 1974; Caplan *et al.*, 1975).

When the concept of stress was first proposed by Selye (1950, 1956), it was as a description of the common physiological effects of a large number of different noxious agents. It is this 'economy' of description that is appealing in the application of the concept to occupational studies. As has been suggested above, a very wide range of events, situations, and physical stimuli can cause problems at work. Many of their effects are similar, although the closely integrated response pattern suggested by Selye must be seriously questioned. In *factor analysis*, instead of describing each factor by listing all the variables that significantly load on it, an arbitrary descriptive label is applied for the sake of economy. Similarly, in discussing occupational problems and the common ground between them it is useful as an economy of description to talk of occupational *stress*. Research will determine how much substance that common ground has and thus the validity of the descriptive label.

Models of stress

The majority of lay individuals have an intuitive and superficial understanding of the term 'stress'. It is widely used in everyday conversation, and has become increasingly popular with the media. However, at a more scientific level, there is a need for precision and clarity of definition; agreement then becomes almost impossible. Many of the authors who have written specifically on the subject of stress present accounts which are sensible when taken singly but inconsistent with one another when the reader attempts to achieve an integrated picture (e.g. Appley and Trumbull, 1967; Lazarus, 1966; McGrath, 1970; Kearns, 1973; Kagan and Levi, 1971; Dodge and Martin, 1970; Janis, 1971; Weitz, 1966; Levine and Scotch, 1970; Moss, 1973; Gray, 1971; Scott and Howard, 1970). The basic differences in approach are discussed below. This discussion has already been presented on several occasions (Cox, 1975, 1978).

Essentially, three main types of definition or model have been advanced. The first type of definition regards stress as a *response*, or a response pattern. It is treated as a dependent variable. The second approach, on the other hand, treats stress as an independent variable for study in that it represents a *stimulus* in the environment external to the person. The third regards stress as a dynamic psychophysiological process, intervening between stimulus and response.

The response-based approach

A simple response-based approach to stress is shown in Figure 7.1. This type of model was given its original impetus in the writing of Hans Selye (1950, 1956). He defined stress as a 'state manifested by a specific syndrome which consists of all the non-specifically induced changes within a biologic system' (Selye, 1950). Selye emphasized the *non-selectivity and non-specificity* of the (physiological) response. On encountering a 'stressor' or 'environmental demand', the organism exhibits a triphasic response, which Selye termed the 'General Adaptation Syndrome' (GAS). The first stage of the syndrome, called the *alarm stage*, is characterized by the general mobilization of resources to meet the demand, centred around a sympathetic–adreno-medullary response. If the demand or stressor persists, this initial stage gives way to the second phase, known as the stage of *resistance*. During this longer phase (assuming demand to be constant), the centre of activity passes from the adrenal medulla to the adrenal cortex. Generally speaking, this phase is characterized by a further increase in catabolic mechanisms which promote or maintain readiness for action and decrease in those which are concerned with reproduction and growth (anabolic). If the demand is prolonged or severe, bodily reserves are eventually depleted, and at this point resistance to demand decreases sharply. The third and final stage, the stage of *exhaustion*, then occurs. During this third phase, the sympathetic–adreno-medullary activity characteristic of the first phase reappears. If demand still persists, death often results. Selye's idea of non-specificity (in terms of demand) has had considerable influence for many years. However, there is now growing evidence that the concept of total non-specificity has been overstated. Mason (1971), for example, has shown that some noxious physical stimuli do not produce the GAS in its entirety. In particular, he has cited physical exercise, fasting, and heat. Lacey (1967) has also argued that the low intercorrelations observed between physiological indices of the GAS refute the concept of a non-specific pattern of response. A similar situation appears to exist with respect to psychological and behavioural indices of stress and between them and physiological correlates (Holtzman and Bitterman, 1956). For example, there is evidence relating to circulating catecholamines and behaviour which suggests that individuals who direct their response 'inwards' in demanding situations show increased excretion of adrenaline whilst those who direct their response 'outwards' show increased noradrenaline (Funkenstein, 1956; Ax, 1953). Such differentiation of response argues against non-specificity. There is, however, much debate over these findings (Frankenhaeuser, 1975).

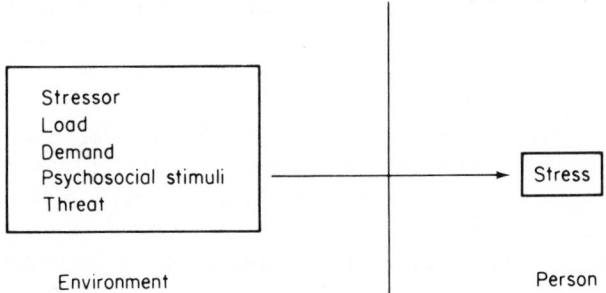

Figure 7.1 A simple response-based approach to stress. Alternative terms for the environmental stress-producing stimuli are shown in the left-hand box

Recently, at the Laboratory for Clinical Stress Research, Levi and Kagan (see, for example, Kagan and Levi, 1971; Levi, 1974) have constructed a theoretical model to describe the way in which social and psychological stimuli cause disease (primarily of a physical nature). The model is largely based upon the ideas of Selye. As well as using the concept of non-specificity, an integral part of the Levi–Kagan model is the relationship between psychosocial stimulation and levels of 'stress' (i.e. disease-producing mechanisms). They suggest that this relationship is best described by a 'U'-shaped function, so that the highest levels of stress are found at the extremes of the stimulation continuum. This introduces the idea that understimulation as well as overstimulation can be a problem. Psychosocial stimuli evoke a physiological response which prepares the person for the physical activity of coping. This response, at least if prolonged, intense or often repeated, causes an increased rate of 'wear and tear' in the individual and eventually results in structural and functional damage. Over a long period of time increased morbidity is the result of this activity. One problem with this model is its failure to spell out the mechanisms by which 'increased wear and tear' occur and translate into disease. The effects of external psychosocial stimuli are conditioned by a number of 'interacting variables'. These may be intrinsic or extrinsic, mental or physical. Another important aspect of the model is what the authors call the 'psychobiological programme'. This is defined as 'the propensity to react in accordance with a certain pattern'. It is partly a reflection of genetic influences and partly a reflection of experience and learning. (Other response-based models can be found in Dohrenwend (1961), Mechanic (1962), and Wolff (1950, 1953).)

Performance degradation as the stress response

McGrath (1970) has discussed a special form of response-based definition, based on performance degradation. This has had particular appeal for many experimental and engineering psychologists. Several difficulties are worth noting. First, many potential stressor variables (say, noise, as one example) produce a wide range of effects: an increase, a decrease or no change in the performance of task. These changes appear conditioned by task, temporal and individual factors. Second, large inter- and intra-individual differences in response exist, so that arriving at an agreed definition of stress as a general change in performance is difficult. Third, performance can be measured in various ways. The situation is invariably complex and the choice of inappropriate or restricted performance criteria for the assessment of stress may result in misleading inferences being made. For example, a slowing of reaction time in a particular experiment may be taken as evidence of stress if only that dependent variable is measured. Additional observations of error scores, however, may indicate the reverse, that performance has improved. What is in fact a change in performance strategy may be misidentified and wrongly labelled as stress. Further discussion of the relationship between occupational stress and work performance can be found in McLean (1974).

Most of these points also apply to the general consideration of stress as a response.

The stimulus-based approach

Whereas the response-based approach regards stress as a *dependent* variable, the stimulus-based approach views stress as an *independent* variable for study, as an objective property of the external environment. In this sense stress is viewed in terms of the stimulus characteristics of environments which are recognized as disturbing or disruptive in some fashion. This approach to stress is often used by ergonomists. A typical stimulus-based approach to stress is shown in Figure 7.2. The basic approach was espoused succinctly by Sir Charles Symonds (1947) when discussing psychological disorders of RAF flying personnel. He wrote . . . 'it should be understood once and for all that (flying) stress is that which happens *to* the man, not that which happens *in* him, it is a set of *causes* not a set of *symptoms*' (our emphasis).

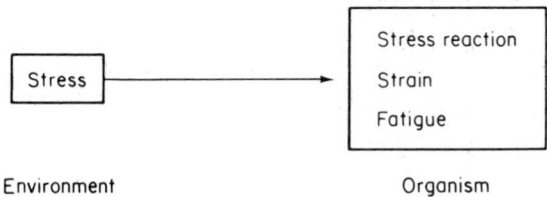

Figure 7.2   A simple stimulus-based approach to stress. Alternative terms for the response to environmental stress are shown in the right-hand box

A particular type of stimulus-based approach has been referred to as the 'engineering analogy'. Here the engineering approach to stress and strain has been borrowed, together with other concepts from the physical sciences. Consider Hookes' law, which describes how loads produce deformation in metals. A load placed upon metal results in a deformation due to internal strain. If the strain produced within the structure of the metal is within the 'elastic limit' of the material, when the load is removed the metal will (eventually) return to its original condition. If, however, the strain passes beyond the elastic limit, then permanent damage will result. The analogy that can be made is that just as metals have different properties, such as different elastic limits, so different individuals have different built-in resistance or 'breaking points'. Up to a point stress can be tolerated, but thereafter permanent damage, either physiological or psychological, results. Individuals may vary in the degree to which they can tolerate stress. Certain individuals may show high stress resistance. The studies describing those individuals eventually selected to become the Mercury astronauts (Korchin and Ruff, 1964; Ruff and Korchin, 1964) illustrate this point. These have been discussed by Cox (1978). Similarly, a recent report by Caplan *et al.* (1975) has shown that 'flexibility', as measured by the California Personality Inventory, has a strong conditioning effect on the relationship between job demands and the subsequent physiological and psychological responses. (See also Caplan, 1971; Kahn *et al.*, 1964.)

A more complex stimulus-based approach has been put forward by Welford (1973). He suggests that stress arises whenever there is a departure from optimum conditions which the person is unable, or not easily able, to correct. Most organisms, including men, appear to have evolved so that they function best under conditions of moderate demand. If a man's performance is less than maximum (leaving aside for the moment the problem of defining what is maximum) this may be, according to Welford, due to either too high or too low a level of demand, both comprising a departure from optimum conditions. This approach is particularly suited to the type of environmental demands described by Broadbent (1963, 1971), Poulton (1970), Wilkinson (1969), and others. These authors relate the effect of environmental demand to performance in terms of psychophysiological concepts such as arousal. This basic proposition has been elaborated upon by McGrath (1970) in the following way: '... the intensity of environmental stimulation (broadly conceived) is curvilinearly related to the degree of felt stress and to the degree of effectiveness of subsequent performance'. A similar approach has been adapted by Frankenhaeuser and her co-workers (for a review see Frankenhaeuser, 1975) as part of a more general psychophysiological and social psychological approach (Frankenhaeuser and Gardell, 1975). Drawing upon the techniques of these two disciplines, they have investigated the impact on the individual of 'underload' and 'overload', which, they suggest, are potent sources of stress in modern industrial society. They draw particular attention to those jobs characterized by *qualitative* underload (i.e. too simple) combined with quantitative overload (i.e. too much to do).

There have been many attempts to produce a list of 'stressful' stimuli, as demanded by a simple stimulus-based approach to stress. For example, Weitz (1970) has described eight different types of situation which have been classed as 'stressful'; these are: speeded information processing, noxious environmental stimuli, perceived threat, disrupted physiological function, isolation and confinement, blocking, group pressure, and, finally, frustration.

General comments

The response- and stimulus-based definitions both represent simple linear models, differing only in what they label as 'stress'. This difference, however, is not trivial in the application of these definitions and leads to very different courses of action. Several general criticisms can be levelled at both types of definition. First, as 'economies of description' they are both too narrow. Second, they are too mechanistic, treating the person as a passive recipient of stress. Third, they give no real status to individual and psychological processes.

This is the overriding drawback with these two types of model. Scott and Howard (1970) have pointed out that '. . . certain stimuli by virtue of their unique meaning to particular individuals may prove problems only to them; other stimuli by virtue of their commonly shared meaning are likely to prove problems to a large number of persons'. This statement implies the mediation of strong individual, as well as situational characteristics.

Psychological models

In psychological models of stress, the roles of perceptual and cognitive characteristics are thought to be crucial in determining individual differences, and are emphasized. A typical psychological approach to stress is shown in Figure 7.3.

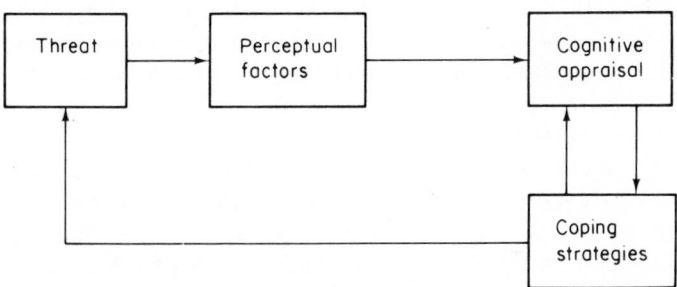

Figure 7.3   A psychological model of stress emphasizing perceptual and cognitive processes

Lazarus (1966, 1976) has presented an important psychological model of stress. He suggests that 'stress occurs when there are demands on the person that tax or exceed his adjustive resources'. An interaction between external demand, the constitutional vulnerability of the person and the adequacy of his defence mechanisms therefore occurs. Lazarus draws particular attention to the person's appraisal of his or her situation and to the role of frustration, conflict, and threat in producing stress. Appraisal refers to the process of assessing or *evaluating* the various elements of the person's situation each against the other. This appraisal process depends upon, amongst other things, learning and past experience. Lazarus emphasizes that appraisal does not necessarily involve conscious mental activity but can occur solely at an unconscious level. Brown (1974) has even suggested that there is evidence that a more accurate testing of reality is achieved by unconscious rather than conscious appraisal. Frustration is regarded as a form of harm which has already occurred to the person by the thwarting or delaying of some important goal. Conflict is the simultaneous presence of two (or possibly more) incompatible action tendencies or goals, and threat is regarded as an imagined or anticipated possible future deprivation of something one values (see also Gross, 1970). The intensity of the threat will depend, apart from the appraisal of its physical characteristics (such as distance, size), upon how well the person feels capable of dealing with the danger or preventing the harm from occurring. If a person feels capable of dealing with the danger or preventing the harm, then threat is minimal. If, however, the person feels helpless and totally unable to master the situation, then threat will be very severe. This again illustrates the interactional nature of Lazarus's approach. The nature of the possible methods by which threat is dealt with is the second major contribution which Lazarus has made to the stress literature. These will be discussed in the next section.

McGrath (1970) has also emphasized that the starting point for any discussion of stress is the nature of the relationship between the individual and his (external) environment. Following from this, he suggests that stress occurs when there is an/ an *imbalance* between environmental demand and the response capability of the individual. Drawing on the work of Lazarus, it is clear that this imbalance is that between the individual's *perceived* demand and *perceived* capability rather than between *objective* levels of these variables. Once the individual attempts to deal (cope) with demand, then a process of secondary appraisal occurs during which the person assesses the degree to which the coping action he has taken, consciously or unconsciously, is meeting demand. The importance of the imbalance between demands and responses has also been pointed out by Mechanic (1962).

A laboratory study which differentiates between objective and perceived levels of demand has been described by Sales (1969). Using an anagram-solving task, Sales manipulated two conditions of objective work load: an underload condition in which subjects were kept waiting for anagrams for approximately 30% of the time, and an overload condition where 35% more anagrams were provided than could be decoded in the time allowed. Subjects were divided into those who reported high levels of *subjective* workload and those who reported low *subjective* workload. Data for two of the dependent variables are of interest here: (a) reported interest in, and enjoyment of the task, and (b) changes in serum cholesterol. On both these variables an interaction between subjective and objective workload occurred. Subjects who received the overload condition but reported low subjective workload and those in the underload condition who reported feeling overloaded both reported high levels of interest and enjoyment in the task. Those given the overload conditions and who felt overloaded and those given the underload conditions and who felt underloaded both had low levels of interest in the task. These latter groups also showed increases in serum cholesterol. The former groups, who showed greater interest and enjoyment in the task, exhibited decreases in serum cholesterol. Sales further analysed these data and found that levels of subjective overload were negatively correlated with scores on the verbal section of the Scholastic Aptitude Test (SAT). Individuals with high levels of verbal ability reported low subjective workload, whilst high subjective workload was reported by those with relatively low verbal ability. A negative correlation between interest in the task and changes in serum cholesterol was also found. These data seem to be in general agreement with the contention that the effects of environmental demands (in this case from a task) are mediated foremost by perceptual factors. One of the primary determinants of the occurrence of psychological (interest in task) and physiological (serum cholesterol) correlates of stress is that of the individual's capability to deal with demand.

Another important element in the determination of stress is the perceived consequences for the individual of failure to meet demand (Sells, 1970; Basowitz *et al.*, 1955). If the consequences of failure to meet demand are minimal for the person then stress will be minimal. Conversely, when failure is important stress will be great. These ideas have recently been expounded by McGrath in the following way (McGrath, 1976): 'There is a potential for stress when an environment situation is perceived as presenting a demand which threatens to exceed the person's capabilities and resources for meeting it, under conditions where he expects a substantial differential in the rewards and costs from meeting the demand versus not meeting it'.

A transactional model of stress

In the previous sections of this chapter some of the existing approaches to the definition of stress have been outlined. It has been argued that some are inadequate in a number of respects, and all differ in their implications for the identification, measurement, and reduction of stress. The remainder of the chapter outlines the authors' model of stress (Cox, 1978), which it is hoped meets the requirements implied in that discussion. The model embodies a number of characteristics, some of which have been mentioned above, that the authors suggest are central to the usefulness of the stress concept. These are oulined below. It describes the dynamic process by which people experience and respond to problems and difficulties.

'Stress' is an individual phenomenon; it is the result of a transaction between the person and his situation. The word *transaction* is used to emphasize the active and adaptive nature of the process. The basis for the model is the relationship between four aspects of the individual and the environment. These are shown schematically in Figure 7.4. The person is continually appraising the demands being made on him by his situation and his ability to meet those demands. Furthermore, his environment may or may not provide the opportunity for him to satisfy his needs. These can be viewed as internally generated demands. (In the work situation, one may refer in particular to those needs outlined by Maslow (1954) and Locke (1976).) The term *demand* is used to denote the request or requirement for physical or mental action, and implies constraints with respect to time. Indeed, the perception of time has been shown to alter with the experience of stress and this may be an important factor in affecting the balance between perceived demand and perceived capability. Furthermore, the person may make use of the rate of change over time of these variables as a means of evaluating his situation. Apart from the quality of the demand, its other aspects which can determine its perceptual representation include distance from the person, frequency of occurrence and rate of change, ambiguity, novelty, and whether or not it is alone or in combination with other demands. For a discussion of these aspects the reader is referred to Appley and Trumbull (1967), Lumsden (1975), Cox (1975), Dodge and Martin (1970), and Broadbent (1963, 1971). Personal resources or capability refer not only to specific and defined skills, which are related to particular

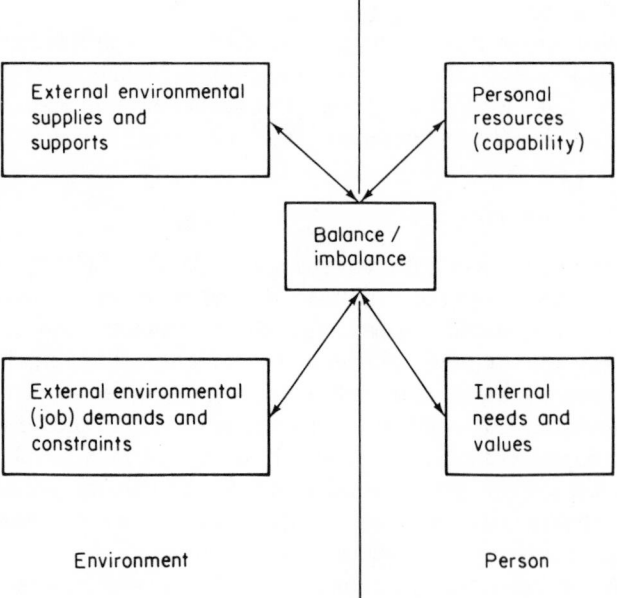

Figure 7.4   The basic components of the transactional model

demands, but also to more general characteristics such as aspects of personality or learnt behaviour patterns.

Other aspects of appraisal and subsequent action are the constraints placed on the person by his situation or his own value systems and, on the other hand, the support offered by others. Constraints may be placed upon the individual limiting his choice of actions. In order to meet or satisfy his needs, an imbalance between demand and capability may be tolerated. In order to earn money to satisfy his physiological and safety needs, the person may accept the underutilization of his skill, because he cannot accept alternative forms of employment or unemployment. Only when the stress arising from this imbalance becomes intolerable may the other demands, his more basic needs, go unfulfilled. The person may resign and subsequently be unemployed. Some individuals may have particularly high levels of internal demand, especially those related to achievement, advancement, and status. They may continually experience an imbalance in demand and capability. For example, a number of authors (Rosenman *et al.*, 1964; Friedman, 1969; Jenkins, 1971) have described a behavioural type, designated 'Type A', in relation to the prevalence of coronary heart disease. Type A individuals, as compared with Type B individuals, are characterized by extreme involvement in their work, competitiveness, a feeling of being under pressure, and above all a need for achievement. Type A individuals often appear prepared to overcommit themselves in terms of the relationship between work demands and their capabilities.

The concept of *imbalance* in the present model can be related to that of 'fit' in the person–environment fit approach of the Michigan school (Van Harrison, 1976; French, 1974; French *et al.*, 1974). Two kinds of fit are emphasized. The first involves the extent to which the person's skills and abilities match the demands and requirements of the job; the second kind of fit relates to the degree to which a person's needs are supplied from his (job) environment. Van Harrison has stated that 'the basic assumption of the theory is that when misfit of either kind threatens an individual's well being, strains will occur such as job dissatisfaction, anxiety, depression and physiological problems'.

An integral part of the present transactional model is the interactions within and between its different levels and stages. Each of these interactions relies upon the concept of feedback mechanisms. They are concerned with maintaining or returning the individual to a state of balance.

Coping involves the use of the person's psychological resources in this way. The first stage of the coping process is concerned with selection of the appropriate responses, and the second with their implementation. Uncertainty concerning appropriateness may constitute a problem in itself. Such uncertainty may have unwanted results, such as anxiety or panic. Thus correct selection and subsequent implementation of coping responses represent a skill in themselves. Lazarus (1966) has divided coping activities into two distinct categories. First, there are behavioural acts called 'direct action tendencies' aimed at reducing external demand directly, and secondly, purely cognitive manoeuvres termed 'defensive reappraisals'. The former, the direct action tendencies, comprise actions such as physical and symbolic attack on the object or person that is perceived to be the agent of harm (persuasion, bargaining, argument). They also include avoidance behaviour, passive behaviour, and activity aimed at increasing capability. Cognitive strategies are aimed at reducing the experience of stress by altering perceptual and cognitive functioning. They include reappraising the content of one's internal needs or altering their rank order of importance, reappraising the consequences of failure to meet external demand or distorting the perception of external demand.

Initiation of coping and its initial effect upon demand leads to a process of secondary appraisal. This may indicate that the imbalance has been reduced, has remained the same, or has increased. The individual must then make a further decision about whether or not to continue using the same coping strategy.

Physiological changes are often associated with the experience of stress. In relation to the present model two points should be emphasized. First, the type of behavioural coping engaged in often determines, at least in part, the pattern of physiological response. A good example of this relationship is the intake-rejection hypothesis of Lacey (1967), or the effect in the work situation of different performance strategies on muscle activity and blood pressure and on attentional mechanisms. The second and related aspect of the physiological response is its effect upon other stages and levels of the model (feedback). Part of the physiological response is based upon the sympathetic adrenal medullary system. It has been argued that this prepares the individual for active behavioural coping in emergency situations. The evocation of this phylogenetically old adaptation pattern may be inappropriate for coping with some of the demands inherent in modern society, since passive or cognitive coping styles may be used. Often physiological resources are thus mobilized but not used, and may 'accumulate' if demand persists. Accumulation of adrenaline, for example, may lead to transient activation of the central nervous system, but prolonged stimulation may give rise to deactivation and a feeling of fatigue (Breggin, 1964). Some physiological responses bring about changes in attentional processes (Kahneman, 1974; Pribram and McGuiness, 1975; Teichner, 1968). For example, Broverman et al. (1975) have hypothesized that the differential biochemical response observed during short and long periods of exposure to stress also effect central attentional processes. These changes in central processing resulting from physiological feedback mechanisms may manifest themselves in changes in the performance of various types of task.

There is much misunderstanding concerning the physiological correlates of stress. Many of the changes which occur are undoubtedly normal homeostatic compensations for the initial response to situations, in terms of the effects of ongoing behaviour or the emergency response. As such they offer no threat to the person. However, certain changes may in some individuals contribute to the aetiology or poor prognosis of disease. A little is known about such an involvement in coronary heart disease, diabetes, ulcers, and cancer (see Cox, 1978; Wayner *et al.*, 1979).

Implications for the reduction of stress

The transactional approach suggests several ways in which the experience of stress can be alleviated and has considerable implications for existing stress reduction techniques. Many traditional methods seek to reduce *stress at source* by reducing or eliminating the external environmental demand along one or more relevant dimensions (e.g. reducing noise level, increasing lighting level). This approach is a characteristic method employed by ergonomics practitioners. It is usual for any remedy to be applied to *groups* of workers. Often, this group approach works reasonably well with physical environmental demands and with man–machine systems design if some flexibility is incorporated. Proponents of the various forms of job redesign argue that increased psychological well-being in the worker population will be produced by adopting these techniques.

The well-publicized job redesign programmes in Scandinavia and the USA have demonstrated the usefulness of such techniques in improving man–job fit. However, in some instances where those methods have been adopted the outcome has not been as successful as one might expect. Proponents of these schemes usually attribute this practical failure to poor design of the new job, but also to poor implementation or mismanagement of the improvement scheme. However, it is suggested here that the assumption that *all* workers in a particular group will benefit from such a scheme is unjustified. Perceived demands, capabilities, and needs, while similar in workers performing the same task, are *not identical*. The result is that job design methods as they are currently practised may improve the psychological (and physical) well-being of many of the workforce but may have no effect or even *reduce* well-being in others. One approach which overcomes this problem has recently been provided by Van Harrison (1976). He advocates a method of stress reduction where *the individual worker becomes responsible for achieving optimum fit* for himself, given an appropriate work situation. The individual worker would be free to modify the structure and content of his job in order to obtain an optimum balance between the demands of his job, his skills and his needs. Although this would be difficult to achieve in practice, it might not be impossible for some jobs providing that the organization's goals and output were maintained (or enhanced).

The transactional model predicts that the *experience of stress* can also be reduced by the alteration of cognitive appraisal. Traditionally this has been achieved by the use of socially acceptable drugs such as alcohol and nicotine (and more recently cannabis) as well as the centrally acting minor tranquillizers. Continuous use of these substances can lead to partial or total dependence and they are not, therefore, entirely satisfactory as stress management tools. Counselling and other psychotherapeutic programmes may be potentially powerful agents for promoting the management of stress, and should be used as an adjunct to chemotherapy. Meichenbaum (1975), for example, has described a stress inoculation skills-training procedure in which the individual is involved in generating self-statements related to appraisal and coping. When used in conjunction with relaxation exercises during stressful experiences, this technique has been shown to be particularly successful in reducing the individual experience of stress. Relaxation alone may mimic some of the useful effects of the minor tranquillizers, such as a reduction in muscle tension.

Two traditional methods of ensuring *man–job fit* are selection and training. These methods are used primarily for middle and high level occupations (clerical, supervisory, professional, management) but not for simple, repetitive shop-floor jobs. However, there is increasing evidence that production-line jobs are subject to as much stress as occupations such as senior management. One possible area of future research therefore would be to examine the different skills needed for technically 'unskilled' work and subsequently to design appropriate selection and training methods. In the industrial context, our definition of skill may need reexamining and extending; 'politically' recognized and specific motor skills obtained through training programmes (e.g. apprentice schemes) are only a part of the range of 'skills' and strategies developed to cope with work. As individuals are promoted or moved within organizations, the demands imposed by the new job increase both quantitatively and qualitatively. An individual's skills may often fail to match the demands imposed by a new job until he has had appropriate training or experience. This is partially supported by the work of Kojima *et al.* (1967), who have shown that new recruits to a production-line job showed high excretion rates of urinary catecholamines when compared with established workers. After several weeks carrying out the job, the excretion rate of the new recruits reduced to the level of the other workers. Possibly, increased familiarization with the job allowed the workers to better manage their personal resources, developing the skills and strategies necessary to cope with the demands imposed. There is a case for paying particular attention to individuals who begin new jobs, for example probationer nurses and teachers. Similarly, extra support could be given to those individuals suffering or recovering from illness. Also worth noting here is that training schemes themselves can be unnecessarily demanding upon those participating (Burrows *et al.*, 1977).

Alteration of the physiological *response to stress* may also be used as a stress management technique. Physiological responses can be either reduced or blocked depending upon situational factors. Some examples are given below. The awareness of tachycardia in some individuals increases the experience of stress (Schachter and Singer, 1962). Drugs such as practolol and propanolol (although the latter is centrally as well as peripherally acting) reduce the experience of stress by reducing tachycardia. In the same vein, Carlsson *et al.* (1972) have demonstrated how the rise in free fatty acids (a cardiac risk factor) brought about by the adrenaline response to stress can be blocked by the prior administration of small quantities of nicotinic acid. In a series of studies (Simpson, Cox and Rothschild, 1974) the present authors and their colleagues have shown that the oral administration of glucose can offset the detrimental effects of moderate intensity noise upon the performance of a number of psycho-motor tasks. Moreover, this effect appears to be related to the level of demand. An unpublished experiment suggests that performance can be maintained by titrating the amount of glucose the subject receives with the amount of noise to which he is exposed.

Discrepancy theories: research in Nottingham

Great emphasis has been placed on definitions and models of stress which treat it as a dynamic process of matching aspects of demand against the person's resources or ability to cope. Different authors add various caveats, but essentially all such approaches propose that it is the discrepancy between demand and capability that is the source of the person's problems and the experience of stress. These approaches have been called 'discrepancy' models or theories for obvious reasons. The leading occupational proponents of such models or theories have been Caplan and his colleagues in Michigan and the present authors in Nottingham. Indeed, the present senior author (Cox) has recently used this approach as the basis of an overview of occupational psychology and as a framework for bringing together its many different aspects (Cox, 1980c; Cox and Brotherton, in preparation).

Both schools have applied their theories in studies on the effects of different occupations on the worker. The Michigan group are perhaps best known for their often quoted report 'Job Demands and Worker Health' (Caplan *et al.*, 1975). In Britain the Nottingham group have been involved in a long-term study of repetitive work. This has been funded by the Medical Research Council and by the US Army Research Institute. Their programme has sought to understand the problems associated with such work and its immediate psychophysiological effects. It has also been concerned with individual differences in psychological response and with their relation to performance strategies.

The programme of research had two aims: one empirical and the other theoretical. The first concerned developing a strategy for studying occupational stress, identifying possible sources of demand and sensitive measures of effect. The second was the development of a model for understanding the effects of the demands associated with repetitive work. Such a model has now been published (Cox, 1980b). It was hoped that fulfilment of these twin aims would allow recommendations for job design and the further study of work processes to be made. The research progressed through an integrated programme of controlled experiments on simulated work systems and field studies in local industry. These have already been summarized (Cox et al., 1979). Briefly, it has been possible to publish a list of some of the demands associated with repetitive work (Cox, 1980c); and to describe at least two patterns of response, one reflecting a 'boredom' effect and the other the interaction between pay and pacing requirements or constraints. This research and an accompanying review of the literature (Cox, 1980b) has suggested the following sources of demand associated with repetitive work:

— The *repetition of a simple motor act*, often associated with an underutilization of skill, or skill potential;
— A predictable and relatively unchanging visual (and auditory) environment, with *restricted levels of stimulation*;
— *Constraints*, often resulting in reduced social contact, in particular high noise levels, high attentional demand and machine pacing, including high rates of pacing;
— *Lack of control*, associated with little responsibility or autonomy, and low participation;
— *Incentive pay schemes*.

These early studies have led to further studies on repetitive work looking in particular at the attentional demands involved and at health processes, and also to studies on machine-minding jobs, which might be prototypical of those created by the current wave of automation.

Concluding remarks

The concept of stress could not be judged to be *essential* for any of the studies reviewed in that it could be replaced by a variety of other concepts. However, it does bring an economy of description to a very wide area of study. Its validity may go beyond that of a mere label, as there is some real common ground between the various problems, processes, and effects that it encompasses. Perhaps this common ground is not as straightforward or extensive as suggested by Selye (1950, 1956), but most would agree it does exist. What is also agreed is that the most adequate models and theories emphasize the dynamic nature of the phenomenon, and treat it as a psychophysiological process. The most recent of such models and theories also emphasize the key role of a discrepancy between demands on the person and his ability to cope. It is along these lines that future approaches to stress will probably develop.

Note

The above is a revised and expanded version of a paper presented to the Third PROMSTRA Seminar, held at the Department of Engineering Production, University of Birmingham, 21–22 September 1976. It develops ideas first presented by Cox (1975, 1978) and Cox and Mackay (1975). The views expressed are those of the authors.

Both authors acknowledge the support of the Medical Research Council during the preparation of the chapter; the first author would also like to acknowledge the support of the Science Research Council.

The authors would like to thank Ann Cooke, Mary Harvey, and June Aylott for typing the manuscript.

References

Appley, M. H., and Trumbull, R. (1967). *Psychological Stress*. Appleton-Century-Crofts, New York.

Ax, A. F. (1953). The physiological differentiation between fear and anger in humans. *Psychosomatic Medicine*, **15**, 433.

Basowitz, H., Persky, H., Korchin, S. J., and Grinker, R. R. (1955). *Anxiety and Stress*. McGraw-Hill, New York.

Breggin, P. R. (1964). The psychophysiology of anxiety. *Journal of Nervous and Mental Diseases*, **139**, 558.

Broadbent, D. E. (1963). Differences and interactions between stress. *Quarterly Journal of Experimental Psychology*, **15**, 205.

Broadbent, D. E. (1971). *Decision and Stress*. Academic Press, London.

Broverman, D. M., Klaiber, E. L., Vogel, W., and Kobayashi, Y. (1975). Short-term versus long-term effects of adrenal hormones on behaviour. *Psychological Bulletin*, **81**, 672–94.

Brown, B. B. (1974). New mind, new body. *Psychology Today*, **8** (3), whole issue.

Burrows, G. C., Cox, T., and Simpson, G. C. (1977). The measurement of stress in a sales training situation. *Journal of Occupational Psychology*, **50**, 45.

Buzzard, R. B. (1973). A practical look at industrial stress. *Occupational Psychology*, **47**, 51.

Caplan, R. D. (1971). Organisational stress and individual strain: a social–psychological study of risk factors in coronary heart disease among administrators, engineers, and scientists. Unpublished Doctoral Dissertation, University of Michigan.

Caplan, R. D., Cobb, S., French, J. P. R., Van Harrison, R., and Pinneau, S. R.

(1975). Job demands and worker health: main effects and occupational differences. US Department of Health, Education and Welfare, Washington, DC.

Carlestam, C., Karlsson, C., and Levi, L. (1973). Stress and disease in response to exposure to noise: a review. Reprinted from Proceedings of the International Congress on Noise as a Public Health Hazard, Dubrovnik, Yugoslavia, by US Environmental Protection Agency.

Carlsson, L. A., Levi, L., and Oro, L. (1972). Stressor-induced changes in plasma lipids and urinary excretion of catecholamines, and their modification by nicontinic acid. In Levi, L. (Ed.), *Stress and Distress in Response to Psychosocial stimuli.* Pergamon, Oxford.

Cooper, C. L., and Marshall, J. (1976). Occupational sources of stress: a review of the literature relating to coronary heart disease and mental ill-health. *Journal of Occupational Psychology*, **49**, 11.

Cox, T. (1975). The nature and management of stress. *New Behaviour*, **2**, 493.

Cox, T. (1978). *Stress*. Macmillan, London.

Cox, T. (1980a). Occupational psychology. In Gilham, W. E. C. (Ed.), *Psychology Today*. English University Press, London.

Cox, T. (1980b). Repetitive work. In Cooper, C. L., and Payne, R. (Eds.), *Current Concerns in Occupational Stress*. Wiley, Chichester.

Cox, T. (1980c). People, work and stress. In *Prevention in Mental Health*. MIND publications, London.

Cox, T., and Brotherton, C. J. (in preparation). *Human Resources at Work*.

Cox, T., and Mackay, C. J. (1975). Stress and the regulation of blood glucose levels. Joint Meeting of the World Health Organisation Psychosocial Centre and the Physiological Psychology Unit, Stockholm University, Stockholm (October).

Cox, T., and Mackay, C. J. (1979). Occupational stress and the quality of working life. In Mackay, C. J., and Cox, T. (Eds.), *Response to Stress*. IPC, Guildford.

Cox, T., Thirlaway, M., Watts, C., Cox, S., and Mackay, C. J. (1979). Job stress: the effects of repetitive work. *Department of Employment Gazette*, December, 1234.

Dodge, D., and Martin, W. (1970). *Social Stress and Chronic Illness: Mortality Patterns in Industrial Society*. University of Notre Dame Press, London.

Dohrenwend, B. P. (1961). The social psychological nature of stress: a framework for causal inquiry. *Journal of Abnormal and Social Psychology*, **62**, 294.

Engstrom, A., Backstrand, G., and Stenram, H. (1971). The human work environment: Swedish experiences, trends and future problems. Royal Ministry for Foreign Affairs and The Royal Ministry of Agriculture, Sweden. Stockholm.

Frankenhaeuser, M. (1975). Sympathetic-adreno-medullary activity, behaviour and the psychosocial environment. In Venables, P. H., and Christie, N. J., (Eds.), *Research in Psychophysiology*. Wiley, London.

Frankenhaeuser, M., and Gardell, B. (1975). Underload and overload in working life: a multidisciplinary approach. Reports from the Department of Psychology, University of Stockholm, No. 460. Stockholm.

Fraser, R. (1947). The incidence of neurosis among factory workers. Industrial Health Research Board of the Medical Research Council Report, No. 90. HMSO, London.

French, J. R. P. (1974). Person–role fit. In McLean, A. (Ed.), *Occupational Stress*. Charles C. Thomas, Springfield, Illinois.

French, J. R. P., Rodgers, W., and Cobb, S. (1974). Adjustment as person–environment fit. In Coehlo, C. V., and Hamburg, D. A. (Eds.), *Coping and Adaptation*. Basic Books, New York.

Friedman, M. (1969). *Pathogenesis of Coronary Artery Disease.* McGraw-Hill, New York.

Funkenstein, D. M. (1956). Norepinephrine-like and epinephrine-like substances in relation to human behaviour. *Journal of Mental Diseases,* **124,** 58.

Gardell, B. (1973). Quality of work and non-work activities and rewards in affluent societies. Reports from the Psychological Laboratories, University of Stockholm, No. 403. Stockholm.

Gray, J. (1971). *The Psychology of Fear and Stress.* Weidenfeld and Nicolson, London.

Gross, E. (1970). Work, organisation and stress. In Levine, S., and Scotch, N. A. (Eds.), *Social Stress.* Aldine, Chicago.

Hackman, R. J. (1970). Tasks and task performance in research on stress. In McGrath, J. E. (Ed.), *Social and Psychological Factors in Stress.* Holt, Rinehart and Winston, New York.

Hinkle, L. E. (1973). The concept of 'stress' in the biological and social sciences. *Science, Medicine and Man,* **1,** 31–48.

Holtzman, W. H., and Bitterman, M. E. (1956). Factorial study of adjustment to stress. *Journal of Abnormal and Social Psychology,* **52,** 179.

Janis, I. L. (1971). *Stress and Frustration.* Harcourt, Brace, Javanovich, New York.

Jenkins, C. D. (1971). Psychologic and social precursors of coronary disease. *New England Journal of Medicine,* **284,** 307.

Kagan, A., and Levi, L. (1971). Adaptation of the psychosocial environment to man's abilities and needs. In *Levi, L.* (Ed.), *Society, Stress and Disease,* Vol. 1. Oxford University Press, London.

Kahn, R. L., Wolfe, D. M., Quinn, R. P., Snoek, J. D., and Rosenthal, R. A. (1964). *Organisational Stress: Studies in Role Conflict and Ambiguity.* Wiley, New York.

Kahneman, D. (1974). *Attention and Effort.* Prentice-Hall, Englewood Cliffs, New Jersey.

Kearns, J. L. (1973). *Stress in Industry.* Priory Press, London.

Kojima, A., Kakizaki, T., and Niyami, Y. (1967). Catecholamine and 17-ketogenic steroid excretion and plasma 11-hydroxy corticosteroids level in new workers with special reference to job adaptation. *Industrial Health (Japan),* **5,** 1.

Korchin, S. J., and Ruff, G. E. (1964). Personality characteristics of the Mercury astronauts. In Grosser, G. H., Wechsler, H., and Greenblatt, M. (Eds.), *The Threat of Impending Disaster.* MIT Press, Cambridge, Mass.

Lacey, J. I. (1967). Somatic response patterning and stress: some revisions of activation theory. In Appley, M. H., and Trumbull, R. (Eds.), *Psychological Stress.* Appleton-Century-Crofts, New York.

Lazarus, R. S. (1966). *Psychological Stress and the Coping Process.* McGraw-Hill, New York.

Lazarus, R. S. (1976). *Patterns of Adjustment.* McGraw-Hill, New York.

Levi, L. (1974). Stress, distress and psychosocial stimuli. In McLean, A. (Ed.), *Occupational Stress.* Charles C. Thomas, Springfield, Illinois.

Levine, S., and Scotch, N. A. (1970). *Social Stress.* Aldine, Chicago.

Locke, E. A. (1976). The nature and causes of job satisfaction. In Dunnette, M. D. (Ed.), *Handbook of Industrial and Organisational Psychology.* Rand-McNally College Publishing, Chicago.

Lumsden, D. P. (1975). Towards a systems model of stress: feedback from an anthropological study of the impact of Ghana's Volta River project. In Sarason,

I. G., and Spielberger, C. D. (Eds.), *Stress and Anxiety*, Vol. 2. Hemisphere, New York.

McGrath, J. E. (1970). *Social and Psychological Factors in Stress*. Holt, Rinehart and Winston, New York.

McGrath, J. E. (1976). Stress and behaviour in organisations. In Dunnette, M. D. (Ed.), *Handbook of Industrial and Organisational Psychology*. Rand-McNally College Publishing, Chicago.

McLean, A. (Ed.) (1974). *Occupational Stress*. Charles C. Thomas, Springfield, Illinois.

Mackay, C. J., and Cox, T. (1975). The nature and study of moderate stress. Joint Meeting of the World Health Organisation Psychosocial Centre and the Physiological Psychology Unit, Stockholm University, Stockholm, October.

Maslow, A. H. (1954). *Toward a Psychology of Being*. Van Nostrand, New York.

Mason, J. W. (1971). A re-evaluation of the concept of 'non-specificity' in stress theory. *Journal of Psychiatric Research*, **8**, 323.

Mechanic, D. (1962). *Students under Stress*. Free Press, Glencoe, Illinois.

Meichenbaum, D. (1975). A self-instructional approach to stress management: a proposal for stress inoculation training. In Speilberger, C. D., and Sarason, J. G. (Eds.), *Stress and Anxiety*, Vol. 1. Hemisphere, New York.

Moss, G. E. (1973). *Illness, Immunity and Social Interaction*. Wiley, Toronto.

Poulton, E. C. (1970). *Environment and Human Efficiency*. Charles C. Thomas, Springfield, Illinois.

Pribram, K. M., and McGuinness, D. (1975). Arousal, activation, and effort in the control of attention. *Psychological Review*, **82**, 116.

Rosenman, R. H., Friedman, M., and Strauss, R. (1964). A predictive study of CHD. *Journal of the American Medical Association*, **189**, 15.

Ruff, G. E., and Korchin, S. J. (1964). Psychological responses of the Mercury astronauts. In Grosser, G. H., Wechsler, H. M., and Greenblatt, M. (Eds.), *The Threat of Impending Disaster*. MIC Press, Cambridge, Mass.

Sales, S. M. (1969). Differences among individuals in attentive, behavioural, biochemical and physiological responses to variations in work load. Unpublished Ph.D. Thesis, University of Michigan.

Schachter, S., and Singer, J. E. (1962). Cognitive, social and physiological determinants of emotional state. *Psychological Review*, **69**, 379.

Scott, R., and Howard, A. (1970). Models of stress. In Levine, S., and Scotch, N. (Eds.), *Social Stress*. Aldine, Chicago.

Sells, S. B. (1970). On the nature of stress. In McGrath, J. C. (Ed.), *Social and Psychological Factors in Stress*. Holt, Rinehart and Winston, New York.

Selye, H. (1950). *Stress*. Acta, Montreal.

Selye, H. (1956). *The Stress of Life*. McGraw-Hill, New York.

Simpson, G. C., Cox, T., and Rothschild, D. R. (1974). The effects of noise stress on blood glucose level and skilled performance. *Ergonomics*, **17**, 481.

Swedish Employers' Confederation (1975). *Job Reform in Sweden*. Swedish Employers' Confederation, Stockholm.

Symonds, Sir C. P. (1947). Use and abuse of the term 'flying stress'. In *Air Ministry, Psychological Disorders in Flying Personnel of the Royal Air Force, Investigated During the War, 1939–1945*. HMSO, London.

Taylor, R. (1974). Stress at work. *New Society*, **30**, 140.

Teichner, W. H. (1968). Interaction of behavioural and physiological stress reactions. *Psychological Review*, **75**, 271.

Theorell, T. (1974). Life events before and after onset of a premature myocardial infarction. In Dohrenwend, B. S., and Dohrenwend, B. P. (Eds.), *Stressful Life Events: Their Nature and Effects*. Wiley, New York.

Van Harrison, R. (1976). Job stress as person environment misfit. A symposium presented at the 84th Annual Convention of the American Psychological Association, Michigan.

Warr, P., and Wall, T. (1975). *Work and Well-Being*. Penguin, Harmondsworth.

Wayner, E., Cox, T., and Mackay, C. J. (1979). Stress, immunity and cancer. In Oborne, D. J., Gruneberg, M. M., and Eiser, J. R. (Eds.), *Research in Psychology and Medicine*, Vol. 1. Academic Press, New York.

Weitz, J. (1966). Stress. Institute for Defense Analysis, Report No. IDA/HQ 66-4622.

Weitz, J. (1970). Psychological research needs on the problems of human stress. In McGrath, J. E. (Ed.), *Social and Psychological Factors in Stress*. Holt, Rinehart and Winston, New York.

Welford, A. T. (1973). Stress and performance. *Ergonomics*, **16,** 567.

Wilensky, H. L. (1960). Work, career and social integration. *International Social Science Journal*, **4,** 543.

Wilkinson, R. T. (1969). Some factors influencing the effect of environmental stressors upon performance. *Psychological Bulletin*, **72,** 260.

Wolff, H. G. (1950). *Life Stress and Bodily Disease*. Williams and Wilkins, Baltimore.

Wolff, H. G. (1953). *Stress and Disease*. Charles C. Thomas, Springfield, Illinois.

Stress, Work Design, and Productivity
Edited by E. N. Corlett and J. Richardson
© 1981 John Wiley & Sons Ltd

Chapter 8

The Causes of Managerial Job Stress: A Research Note on Methods and Initial Findings

Judi Marshall
University of Bath, UK
and
Cary L. Cooper
University of Manchester Institute of Science and Technology, UK

The initial phase in this research programme was a small-scale study in a large company with several sites to investigate the effects of relocation on their managers and their families. Managers and their wives were interviewed about their past careers and current jobs. It soon became apparent that rather than causing its own distinct problems, 'moving' served to bring to the fore those which were already inherent in the manager's job and life-style. These were, briefly:

1. Work overload, and a consequent lack of spare problem-solving capacity;
2. Time-management problems—especially as many could not get their work done during working hours and doing work at home had become a common 'leisure activity';
3. Strains of travel;
4. Work–home conflicts—accentuated in 'special' situations when the company dealt with the manager alone and ignored the fact that he was part of a larger decision-making unit, the family;
5. Company norms that work performance should be maintained at a high level no matter what pressures the manager was under.

For the second phase, therefore, the focus of study was shifted from the specific topic of managerial mobility to the more generic problem area of

managerial work stress. Figure 8.1 summarizes the research plan for this
investigation, which was empirically as opposed to interview based and
involved extensive questionnaire data collection.

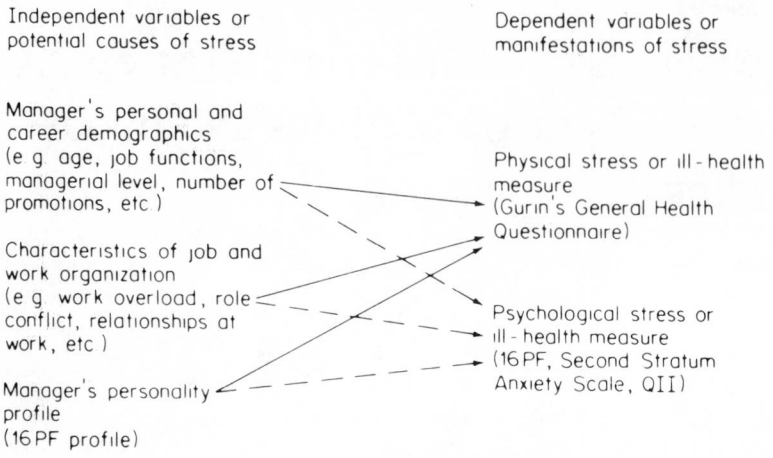

Figure 8.1   Research plan for quantitative data collection phase on causes of managerial job
stress

   On the left-hand side of the figure the factors which were thought to be
potential *causes* of stress are shown. These were derived from interview and
literature data and cover aspects of the individual—his demographics and
personality profile—as well as environmental factors—job and
organizational characteristics. The main purpose of the statistical analysis to
be reported here was to see if stress could be predicted from these
independent variables. Two measures of the *manifestations* of stress were,
therefore, included as dependent variables—one of these focusing on
physical and the other on psychological ill-health symptoms. If predictive
relationships can be established, the hypothesis that some of these
independent variables are causes of stress is supported and we can go on to
consider what these are and explore the varying patterns which emerge in
different circumstances.
   In this chapter, then, we shall briefly describe the research tools used and a
section of the anlysis of results which links emotional and physical ill-health
to the individual's personal and caneer demographics, job and organizational
characteristics, and his personality make-up. (For a full account of the study
see Marshall and Cooper, 1979.)

Independent variables

Questionnaire 1 : Demographic information

This covered *individual* demographics—such as age and life stage—and *career* demographics—job function, number of promotions, etc.—and was tested and refined during the mobility study.

Questionnaire 2 : Job and organizational characteristics

This was also designed specifically for this research and will be described in some detail as it has several unique features and is the main source of potential job and organizational stressors and satisfiers in the study.

From the literature and interview data a classification of potential sources of job stress for managers was drawn up—see Table 8.1 (Cooper and Marshall, 1975).

The Job and Organizational Characteristics Questionnaire was designed to cover the six 'external' (i.e. environment-based) categories. On examining the job areas thus identified, we find that:

1. Almost everything in the work situation is at some time, or by someone, identified as a cause of stress;

2. Frequently both a situation and its direct opposite can cause stress, for example overwork or underwork, too many decisions to make or too few;

3. Many of the factors quoted have been identified in other studies as direct or indirect sources of job satisfaction, for example a 'poorly defined task', whilst causing anxiety, can also provide scope for the worker to use his initiative and gain greater satisfaction from a job well done.

It was therefore decided to treat each job factor as a potential source of satisfaction as well as stress in order to arrive at a comprehensive view of the manager's job situation. A pool of items most of which could be scored either negatively (as a pressure) or positively (as a satisfaction) was developed. After a pilot study of the questionnaire, the pool was reduced to 61 questions which could be rated (on a five-point Likert-type scale) as either a pressure or a satisfaction or as both a pressure and satisfaction. A segment of the questionnaire is shown in Table 8.2 to illustrate this format.

To reduce the questionnaire items to a manageable number for analysis and to test out the categorization on which it was based, scores on the questionnaire's pressure and satisfaction subscales were factor analysed

Table 8.1 Sources of stress at work

1. *Intrinsic to job*
    Too much work Qualitative
                     Quantitative
    Too little work
    Time pressures/deadlines
    Poor physical working conditions
    Mistakes
    Too many decisions

2. *Role in organization*
    Role ambiguity
    Role conflict
    Too little responsibility
    No participation in decision-making
    Responsibility for people
    Responsibility for things
    Lack of managerial support
    Increasing standards of acceptable performance
    Organizational boundaries (internal and external)

3. *Relations within organization*
    Poor relations with boss
    Poor relations with colleagues and subordinates
    Difficulties in delegating responsibility
    Personality conflicts

4. *Career development*
    Overpromotion
    Underpromotion
    Lack of job security
    Fear of redundancy/retirement
    Fear of obsolescence
    Thwarted ambition
    Sense of being trapped

5. *Organizational structure and climate*
    Restrictions on behaviour (e.g. budgets)
    Lack of effective consultation and communication
    Uncertainty about what is happening
    No sense of belonging
    Loss of identity
    Office politics

6. *Organization interface with outside*
    Divided loyalties (company *vs.* own interests)
    Conflicts with family demands

7. *Intrinsic to individual*
    Personality (tolerance for ambiguity, stable self-concept, etc.)
    Inability to cope with change
    Declining abilities
    Lack of insight into own motivation and stress
    Ill-equipped to deal with interpersonal problems
    Fear of moving out of area of expertise

Table 8.2   Extract from Questionnaire 2: job characteristics

	Pressure						Satisfaction					
1. I think it likely I shall be asked to retire early	5	4	3	2	1	0	1	2	3	4	5	NA
2. I sometimes have to work long hours and/or take work home to get things done	5	4	3	2	1	0	1	2	3	4	5	NA
3. I sometimes find myself passing on orders I don't agree with	5	4	3	2	1	0	1	2	3	4	5	NA
4. I have probably reached my career ceiling	5	4	3	2	1	0	1	2	3	4	5	NA
5. Later in my career I might well be asked to change to a completely different type of work	5	4	3	2	1	0	1	2	3	4	5	NA
6. I am often faced with a choice between family and work demands	5	4	3	2	1	0	1	2	3	4	5	NA
7. I do not have to take many decisions on my own	5	4	3	2	1	0	1	2	3	4	5	NA
8. My physical working conditions are. ...	5	4	3	2	1	0	1	2	3	4	5	NA
9. The time pressures and deadlines in my job are. ...	5	4	3	2	1	0	1	2	3	4	5	NA
10. I find managing people. ...	5	4	3	2	1	0	1	2	3	4	5	NA
11. My contacts with people in other departments are. ...	5	4	3	2	1	0	1	2	3	4	5	NA
12. My rate of promotion is. ...	5	4	3	2	1	0	1	2	3	4	5	NA
13. On the whole the content of the job I do is. ...	5	4	3	2	1	0	1	2	3	4	5	NA

separately. Ten pressure and eight satisfaction factors were derived (and respondents' scores on each calculated); these are shown in Table 8.3 in their descending order of magnitude for the sample as a whole, and are the 'job factors' on which the analysis reported below is largely based.

Questionnaire 3: Manager personality profile

Cattell's 16PF (Form C) was chosen as a measure of personality as it is a comprehensive, relatively short, profile-oriented scale which has been used extensively on non-clinical populations.

Dependent variables

Selecting a stress measure is tantamount to defining stress for the purposes of the planned research study and must therefore be done with some care. A variety of stress measures are available to the researcher. These are pitched at three levels of symptomatology—the psychological, physical and

Table 8.3 Job factors derived from factor analysis: in descending order of magnitude for the sample studied

Pressures	Satisfactions	Description
	Relationships	Predominantly peer relationships
	Recognition	Career success in terms of pay, promotion, responsibility, etc.
	Managing people	Relationships with subordinates
Overload		Having more work than one can cope with
	Challenging job	Job demands—especially those for time and activity
Swamped in big company		Company's slowness in reacting to change, poor communications, and conflict from implementing policies one doesn't agree with
Job security		Fear of job loss
Career development		Concern about promotion, pay, and career prospects
Uncertainty		Uncertainty and a lack of control over one's own life
Lack of autonomy		Alienation and dissatisfaction with organizational role
	Conflict	Striving against both colleagues and company—but from a secure position
Underload		Thwarted ambition and work underload
	Autonomy	Independence
	Faith in company	Trusts company and able to perform job easily
Job challenges		Job demands—especially those on one's time and of dealing with people
Relationships		Pressure from relationships with colleagues and subordinates
	Belonging to large company	Security of large company membership
No authority		Lack of responsibility and involvement

behavioural—and range from the objective (blood pressure, serum cholesterol levels, number of hours worked) to the purely subject-defined. All have merits in particular contexts. In our case practical limitations on research design confined us to the use of self-completion questionnaires. It was decided to include two stress measures—one for psychological and one for physical symptoms.

A psychological stress measure

Anxiety being the primary psychological symptom of stress, the 16PF second-stratum factor was selected as a measure of emotional ill-health. It is reasonably reliable, has been extensively validated, and has been used in other stress-orientated studies.

A physical stress measure

Practical limitations meant that 'objective' data sources—medical records, etc.—were not available to us. After some consideration and the piloting of two widely used questionnaires linked to stress research, of the symptom check-list type, it was decided to use the Gurin Psychosomatic Symptom List

Table 8.4   Extract from Questionnaire 3: general health

	A lot	Quite often	Occasionally	Never
Do you find it difficult to get up in the morning?				
Does ill health ever affect the amount of work you do?				
Are you ever bothered by shortness of breath when you were not exercising or working hard?				
Have you ever been bothered by your heart beating hard?				
Do you ever smoke, drink or eat more than you should?				
Do you ever have spells of dizziness?				
Are you ever bothered by nightmares?				
Do your muscles ever tremble enough to bother you (e.g. hands tremble, eyes twitch)?				
Do you ever feel mentally exhausted and have difficulty in concentrating or thinking clearly?				
Are you troubled by your hands sweating so that you feel damp and clammy?				
Have there ever been times when you couldn't take care of things because you just couldn't get going?				

(Gurin, *et al.*, 1960)—with some slight modifications for a British population—as a measure of physical ill-health. This scale has the advantages of being short and relatively non-clinical and its use in social research has been well documented. Examples of scale items are shown in Table 8.4.

Respondents are asked to score their frequency of symptoms on a scale 'a lot', 'quite often', 'occasionally', 'never', and an overall 'health' score is calculated (in the interests of consistency, this has been reversed to give an 'ill-health' score in the analyses which follow). In our case, as the emphasis was on current physical state, the respondent was asked to report his symptoms for the last three months only. As the scale is highly susceptible to faking, we must consider respondents' scores as a conservative estimate of symptomatology; this does not, however detract from their relative value. (Examination of 'motivational distortion' scores on the 16PF—the 'lie detection scale'—suggests that the sample were, in fact, relatively truthful, showing little social desirability distortion.) The sample showed a wider range of scores on this scale than on the others used in the pilot study.

The 'package' of questionnaires was for the most part distributed personally and the response rate was high; approximately 90% of those contacted took part in the study.

The sample

As a final prelude to the results, a word about the sample. This consisted of just under 200 senior managers from a large company; respondents ranged from 27 to 60 years old, the typical manager being in middle age, married, with a growing family; five job functions were represented—research, production, service, marketing, and engineering.

Results

The statistical analysis technique of stepwise multiple regression (an SPSS subprogramme) was used to relate independent to dependent variables in a manner which takes interactive effects into account. This statistical technique is a method of achieving the best linear prediction equation between a given set of independent variables and the dependent variables in question. Variables are incorporated in the equation in their order of significance, thus allowing the researcher a margin of discretion as regards the cut-off point to choose. In this case the following criteria were used:

(a)   That the overall F for the equation was significant;
(b)   That the partial regression coefficient for the individual independent variable being added was at a statistically significant or approaching

significance level. Below this point not only is the coefficient insignificant but also the amount of variance contributed by each additional variable (R^2 change) is very small.

Whilst the variables included here were chosen as being likely 'early contributors' to the stress sequence, correlational analysis does have certain fundamental drawbacks as a basis for causal inference. It must therefore be borne in mind that significantly related factors may be subsidiary symptoms or consequences rather than causes of stress. As our main aim was description rather than hypothesis-testing, attention to the methodological weaknesses of this technique will be suspended for the time being.

Here, regression equations for both dependent variables—anxiety and physical ill health—will be discussed for the sample as a whole and for function subsamples. Of particularly significant note is the greatly increased predictive power of equations based on meaningful function subgroups as opposed to those derived from scores of the total sample.

Regression analysis for the total sample

The pool of independent variables (individual and career demographics, characteristics of the individual's job, and individual personality profiles) were regressed separately to each of the two dependent stress measures. As the anxiety score was calculated wholly from the first-stratum 16PF factors, the six of these which explained the greater part of variations in its score (89% in this case) were omitted from psychological ill-health analyses. (These were 'tense', 'apprehensive', 'affected by feelings', 'suspicious', 'undisciplined self-conflict', and 'shy'.) We should therefore expect poorer predictive power in these than in the physical ill-health equations as significant contributory, but potentially confounding variables have been purposely excluded in order to discover the effects of other factors. Table 8.5 summarizes the regression equation findings.

For the sample as a whole, four variables explain 23.8% of the anxiety score variance. The emergence of two 16PF dimensions (in view of the score's initial derivation) shows that personality is the main cause of high anxiety. We see that it is the 'calculating' but possibly less able individual, in a job characterized by overload and a lack of autonomy, who is 'at risk' of showing psychological stress symptoms. The contribution of 'lack of autonomy' is at first sight surprising in view of the seniority of the sample studied and reveals that dissatisfaction with worker participation can be a frustration even at these levels.

The overall equation explaining physical ill-health bears out the dominant role of personality factors—four explaining 32.5% of the score variation—to the exclusion of job and demographic variables. Two of these factors

Table 8.5   Multiple regression analysis: to derive stress-prediction equations for the sample as a whole and by job function. Variables in left-hand column associated with *higher anxiety scores*. Variables in right-hand column associated with *poorer physical health*

Total sample

23.8%	Shrewd*	Affected by feelings*	32.5%
	Overload P	Tenderminded*	
	Less intelligent*	Tense*	
	Lack of autonomy P	Humble*	

Anxiety (A) + physical ill-health (P.I.) scores were significantly related ($r = 0.47$, $p = 0.001$) but overlap is not perfect.

Research function

70%	Assertive*	Satisfaction outside work	82%
	Tough-minded*	Managing people S	
	Managing people S	Belonging to large co. S	
	Fewer relocations	Less intelligent*	
	Smoking	No. of secondments	
	Spending less time on work	Self-sufficient*	
	Older	Controlled*	
	Self-sufficient*		

A + P.I. correlate only 0.23 (n.s.), i.e. individuals in this group do not tend to show both types of symptoms.

Production

80%	Job security P	Affected by feelings*	71%
	No. of secondments	Humble*	
	Overload P	Lower managerial level	
	Reserved*	Older	
	Belonging to large co. S	Non-smoker	
	Challenging job S	Wife less likely to work	
	Lack of autonomy P	Less intelligent*	
	group dependent*		

A + P.I. were not significantly correlated ($r = 0.32$).

Service depts.

59%	Shrewd*	Tense*	72%
	Lack of autonomy P	Autonomy S	
	Increasing time since last relocation	Humble*	
		Affected by feelings*	
	Decr. time since last promotion	Wife less likely to work	
	Belonging to large co. S	No authority P	

For this group there is a high correspondence between psychological and physical symptoms ($r = 0.67$, $p = 0.001$).

Table 8.5: (contd.)

Marketing

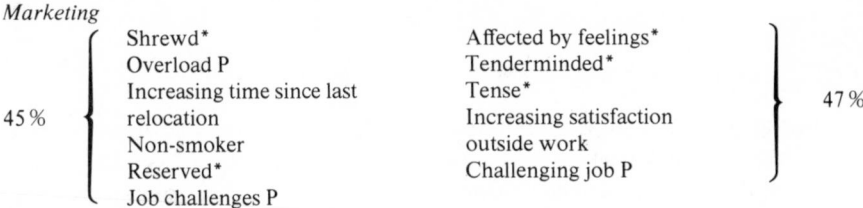

45 %

Shrewd*
Overload P
Increasing time since last
relocation
Non-smoker
Reserved*
Job challenges P

Affected by feelings*
Tenderminded*
Tense*
Increasing satisfaction
outside work
Challenging job P

47 %

A + P.I. are highly significantly correlated ($r = 0.56$, $p = 0.001$).

Engineering

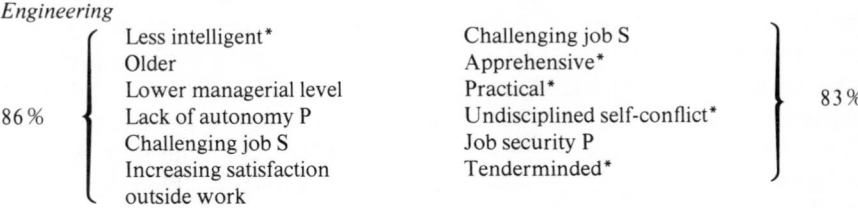

86 %

Less intelligent*
Older
Lower managerial level
Lack of autonomy P
Challenging job S
Increasing satisfaction
outside work

Challenging job S
Apprehensive*
Practical*
Undisciplined self-conflict*
Job security P
Tenderminded*

83 %

A + P.I. are highly significantly correlated ($r = 0.65$, $p = 0.003$).

S, satisfaction; P, pressure; * 16PF factor.
For each analysis the percentage variation in the dependent variable score (anxiety or physical ill health)
explained by the regression equation is given. This provides some indication of the worth of the achieved
solution—in social science research it is unusual to achieve above 50–60 % explanation. Higher percentage
explanations are achieved, somewhat artificially, if the available sample is small.
Function group samples range in size from 20 to 64.

('tenderminded' and 'humble') are, however, relatively unrelated to anxiety,
suggesting that the overlap between our two stress measures is not perfect.
This is borne out by the relatively low statistically significant correlation (also
shown in Table 8.5) between them ($r = 0.47$).

Overall, then, we find that the psychological and physical stress measures
used are closely but not perfectly related and, examined on a sample-wide
basis, are almost wholly personality-determined. The failure of job and
organizational characteristics to play a significant part in predicting stress
was somewhat surprising and the possibility that this was due to their
systematic variation across the population was considered. The fact that
scores on seven of the 18 job factors differed significantly between job
functions provided some support for this view and encouraged us to break
down the data by function. Multiple regression analyses were, therefore,
conducted within function subgroups which elicited equations of much
greater predictive power—many achieving explanations of 70–80 % of the
variance. These are also shown in Table 8.5 and profiles of the managers 'at
risk' in each functional area are described in detail below. The aim will be to

capture the gestalt in each case rather than describe the findings in meticulous detail.

Managers in research departments

For this group, the independent variables or potential stressors linked to each of the dependent variables (psychological and physical stress) are closely similar and will be described as one. The researcher 'at risk' of experiencing stress at work is 'independent' but 'emotionally less stable', 'older' but 'less widely experienced'. His job appears to require characteristics—the ability to manage people and to feel at ease working in a large company—which are in conflict with his personality and past experience. His diversion of time and commitment to activities outside work put him under additional pressure. This picture is one of a manager whose needs and abilities do not match the demands inherent in his job—we might say that there is 'poor P:E fit'.

Anxiety and physical ill-health scores do not correlate significantly, suggesting that the 'stressed' research manager will show either one or the other but not both types of stress symptoms (i.e. physical or psychological).

Managers on production sites

This is the only function for which psychological and physical risk profiles are sufficiently distinct to warrant separate description. Looking at the psychological stress regression equation, we find that for managers in production *high anxiety* is linked to job and organizational characteristics of the manager's work and not his demographic or personality predispositions. Three themes emerge from the regression analyses:

(a)   A lack of job security;
(b)   Occupying an overloading and challenging job;
(c)   Feelings of frustration at belonging to a large company and the lack of autonomy this involves.

The anxious production manager appears to derive satisfaction from his immediate work context (i.e. he shows good P:E fit) but to have poor relationships with higher management—in this case geographically distant, centralized decision-makers.

As the psychological and physical ill-health are not significantly correlated ($r = 0.32$) and the two 'risk profiles' show no overtly common trends, we cannot assume that these managers will show both syndromes. *Physical ill-health* (in the second regression analysis) is, in fact, related to lack of achievement—the older manager at a relatively low level of seniority—and personality characteristics which suggest a tendency to react badly to failure. The manager's wife is unlikely to work full-time and perhaps she too relies on his success for life satisfaction, putting him under added pressure.

We can, then, identify two types of individual at risk in this group, the suggestion being that they show distinct stress symptoms—anxiety with an active, and physical ill-health with a passive situation and reaction.

Managers in service departments

Managers from accounts, personnel, management services, etc., were included in this analysis. The two 'stress configurations' are very similar, with personality variables playing a vital part in both analyses. The implication is that anxiety ('tense' and 'affected by feelings') is closely associated, possibly causally, with physical ill-health. (The two scores are very significant correlated, $r = 0.67$, $p < 0.001$.) The service department employee is, then, unhappy in the job to which he has been newly promoted (interview data suggest that this is a common but usually a temporary phenomenon) because he is overleaded and because it does not live up to his expectations for increased personal autonomy. In fact, he shows signs of feeling trapped by his growing involvement in (and dependence on) the company.

Managers in marketing and sales departments

Whilst the job elements emphasized differ slightly (in that the lack of power and feeling trapped themes do not emerge), the profile of managers at risk in marketing and sales departments bears considerable resemblance to that of those in service departments.

Again, the physical ill-health regressions are dominated by anxious personality traits, and the appearance of factor N ('shrewd, calculating, worldly, penetrating') as the most important contributor to anxiety suggests ambition and career striving; again, also, the close association of the two dependent stress measures suggested by common themes in the 'profiles' is borne out by their significantly high intercorrelation ($r = 0.56$, $p = 0.001$). The marketeer at risk, then, wants advancement but is currently overpowered, both quantitatively and qualitatively. He has, perhaps, been overpromoted, as there is no evidence to suggest that, like the service employee, his job demands have recently peaked; in fact, distraction of energies outside work is a more likely contributory cause.

Profiles of the service and marketing managers under stress are dominated by career development problems, which suggests that ambitious striving has resulted in a, hopefully temporary, inability to cope with work demands.

Managers in engineering departments

The personality and job factors appearing in connection with stress symptoms for engineers are markedly different from those of all four other functions (amongst which, particularly as regards personality, there are similarities); whilst the two regression equations are elaborated in terms of different variables, they show a common theme which suggests that they describe overlapping syndromes. The statistically significant correlation of anxiety with physical ill-health scores ($r = 0.65$, $p = 0.003$) supports this interpretation.

The engineer 'at risk' shows 'conventional', 'dependent', 'practical' but 'worrying' tendencies. The job factors about which he is concerned are *security*, rather than growth-oriented, and it would appear that he feels he has been 'passed over'. He is even unhappy about the challenges in his job—perhaps because he does not have the power and authority to perform as he would like. In this case, seeking satisfaction outside work may well be an attempt to compensate for reduced job involvement.

In this chapter we have outlined an analysis technique by which we can identify and describe those personality and job factors associated with stress symptoms for particular population subgroups. At a conceptual level, too, the above analyses have revealed certain things:

1. That statistical analysis (in this case multiple regression) is a more powerful tool if meaningful subgroups are isolated;
2. That job as well as personality factors appear to contribute to stress;
3. That whilst the P:E fit model is adequate in some instances—e.g. for the research manager—in others (e.g. production managers) a wider organizational context must be investigated to identify the main causes of stress;
4. That the relationship between psychological and physical stress symptoms varies with the person:environment combination concerned.

References

Cooper, C. L., and Marshall, J. (1975). Stress and pressure within organisations. *Management Decision*, **13**(5), 292–304.

Gurin, G., Veroff, J., and Feld, S. (1960). *Americans View Their Mental Health*. Basic Books, New York.

Marshall, J., and Cooper, C. L. (1979). *Executives Under Pressure*. Macmillan, London.

Stress, Work Design, and Productivity
Edited by E. N. Corlett and J. Richardson
© 1981 John Wiley & Sons Ltd

Chapter 9

The Extent and Significance of Stress Symptoms in Industry—with Examples from the Steel Industry

R. J. Tinning
74 Radnor Rd., Harrow, U.K.
and
W. B. Spry
Bryn-y-Neuadd Hospital; Llanfairfechan, Wales

It is intended in this chapter to discuss and consider the following:

1. How extensive are stress symptoms in industry?
2. Which aspect of the total problem of stress is the legitimate concern of industry?
3. How can that aspect of the total problem that is industry's main concern be measured (utilizing a specific case study to illustrate what is involved)?
4. What percentage of the total problem of stress in industry might realistically be influenced from within industry (as indicated by the above case study and by previous social survey studies of stress illnesses)?
5. Do transitory symptoms of stress evolve into chronic conditions, and if so, how?

The extent of stress symptoms in industry

Evidence that symptoms of stress and emotional illness are extremely widespread in working populations can most easily be seen by reference to the familiar and often quoted sickness absence figures for stress-related disorders.

When the National Association of Mental Health first produced its report on Stress at Work in 1971, $36\frac{1}{2}$ million working days were lost in a year

129

through poor mental health and stress-related symptoms. The most recent figures reveal that the annual total is now over 40 million days.

Even this does not give the complete medical picture as these figures do not include absences of less than three days. Nor do they include all illnesses in which stress or tension plays a part. It has been said, for instance, that 40–50% of all medical complaints brought to a clinic may be emotional in origin (Gwynne Jones, 1966). In addition, many physical illnesses have accompanying emotional conditions that delay recovery, or that extend absences originally resulting from organic causes (Collins, 1961).

Sickness absence figures for specific stress illnesses are, however, only the most obvious and by no means the largest part of the problem. To ascertain the full picture it is necessary to make reference to the findings of the intensive social surveys that have been carried out in recent years, aimed at determining the level of stress symptoms or of minor forms of psychiatric ill-health amongst those not receiving medical attention, as well as those who have in some way come to the attention of medical agencies.

Several such research studies have shown that 25–30% of total populations exhibit clinical signs of poor psychological health of a sufficient degree to disturb their everyday lives, and have also shown that one-third of these may be seriously impaired in their personal, social or working effectiveness.

As only one in three of those with symptoms are treated by doctors (or by other professionals), the 40 million days of absence for specific emotional illnesses can clearly be seen as the smaller part of the problem. Yet the hidden and larger portion undoubtedly has adverse effects both on the individuals concerned and on the industry in which they function.

Such effects, it has often been suggested, may be absenteeism, accidents, interpersonal or group conflict, frequent illness (both physical and psychological), working ineffectiveness, and extreme unhappiness (Collins, 1961; Harrington, 1962; Ferguson et al., 1965). Perhaps even more important, the full potential of individuals who suffer from such conditions may never be fully utilized, to their own and industry's cost.

What aspect of the total problem is the legitimate concern of industry?

The evidence of absence statistics and previous research provides some idea of how extensive the stress problem is, indicates some of the consequences to individuals and to industry, and suggests others. While, however, there tends to be a general acceptance that 'stress' symptoms *are* widespread in modern society, doubts still exist as to the exact extent of such symptoms in working populations and also as to their true nature and long-term significance. For instance, although the figures of 25–30% for the proportion of the working population with minor emotional conditions and 5–10% for those having more severe states have been quoted with some consistency (Ferguson et al.,

1965), some general population surveys have found a much higher prevalence rate for psychological illness, and the variation between the lowest and highest figures is considerable.

Prevalence levels reported in survey findings have in fact varied between 8 and over 60% in general populations (Dohrenwend and Dohrenwend, 1965). It is very obvious that these could not all have been measuring the same thing. The variations are the results of different concepts or definitions of mental health or illness, and of different techniques and different intensities of research. In general, the more intensive the survey and the more direct the means of obtaining the information, the higher the rates of emotional disturbance disclosed. The social surveys with the highest figures are in effect a ragbag of conditions including psychotic disorders, neurosis, epilepsy, mental deficiency, and personality disorders; psychosomatic symptoms and transitory anxiety reactions in basically healthy people have also been included. They have therefore taken into account all conditions short of an idealistic state of perfect mental health. Some surveys have equated short-term anxiety or stress symptoms found at one point of time with recurrent symptoms in pathological conditions (Dohrenwend and Dohrenwend, 1965). This assumption throws doubts on some survey findings, as Dohrenwend and Dohrenwend have pointed out.

It is clear that not all the conditions listed above are equally problematical to industry, nor can industry have a direct responsibility for all such states. The more severe conditions will be treated medically, or are primarily a question of careful placement. Indeed, some rather extreme personality types or neurotic traits can even aid individuals in certain tasks. Individuals with personality disorders or neurotic traits do not necessarily display symptoms of stress if they are appropriately placed, nor if they are working within their normal coping capacities (Markowe and Barber, 1953). On the other hand, mentally healthy people placed under severe stress may display symptoms of mental ill-health (Taylor et al., 1964). Whether such symptoms always diminish or cease when external stresses are removed is unclear, but there is evidence to suggest that in some circumstances a proneness to repeat such symptoms under future similar stresses may result (Leighton et al., 1963).

There are situations where it may be desirable to attempt the obtaining of a total picture of all the stress-related conditions, illnesses, and symptoms in a particular population. Prevalence studies carried out in the United States in the sixties (Leighton, et al., 1963; Srole et al., 1962) took this approach, and their findings indicated that less than 20% of the populations examined were entirely free of symptoms of emotional disturbance. There is a danger, however, that when the results of such extensive and all-embracing surveys (inclusive of all conditions, short-term and chronic) are contemplated, the sheer apparent amount of emotional ill-health and stress states disclosed may induce a feeling that the whole problem is just too large to be influenced. Many industrial managers initially confronted with the results of such surveys may understandably feel that it is outside their power, or indeed anyone's within industry, to affect in any realistic way, or to any worthwhile extent. They may as a consequence be extremely reluctant to cooperate in research into a problem that they see too difficult to influence.

Before carrying out research (or action) in this area on a specific industrial population, it is therefore essential to establish the aspect of mental health, mental illness or stress with which industry should legitimately and directly be concerned. To answer this question it is necessary to enter into a brief discussion on the concepts of mental health and illness and their relationship to stress.

Mental health

A considerable time could be spent in discussing the various models of mental health, mental illness and stress, both social and medical, that have been theoretically proposed or informally arrived at for practical needs, but for the purpose of carrying out mental health or stress research in industry two models have particular relevance, those of Jahoda and Klein.

Jahoda (1958) has suggested that much of the confusion in the area of mental health stems from failure to establish whether one is talking about mental health as a lasting attribute of a person or as a temporary attribute of functioning. Therefore she indicates two ways of looking at mental health, as:

1. A relatively constant and enduring function of personality—leading to predictable differences in behaviour and feeling depending on the stresses and strains of the situation in which a person finds himself.
2. A momentary function of personality and situation.

To look at mental health in the first way will lead to a classification of *individuals* as more or less healthy; to look at it in the second way will lead to a classification of *actions* as more or less healthy. Jahoda illustrates the importance of this distinction by an example concerning physical health. If a man with a bad cold is considered in relation to the first classification he would be regarded as healthy, but in relation to the second he is sick. Both suggestions are justifiable and useful, states Jahoda, but confusion will result if either diagnosis is made in the wrong context—that is, if he is regarded as a permanently sick person or as one who is functioning healthily.

Much of the descriptive conflict in the field of mental health, suggests Jahoda, stems from the failure to establish which category of mental health is being talked about or measured.

On rather similar lines Klein (1960) also distinguishes between long- and short-term mental health, but for greater clarity uses three categories. These categories are Soundness, Stability, and Well-being.

Soundness refers to the level of integration of the general, more enduring personality structure, where the individual's early environment is important in personality development and long-term mental health.

Stability refers to the ability of the individual to cope with environmental stresses or illness-producing agents, whilst maintaining a state of emotional well-being. Klein's 'stability' can be either general, the capacity to maintain 'well-being' in regard to a wide range of stress, or 'specific', the ability to withstand particular stresses. Both general and specific stability *may* exist in individuals of only moderate soundness.

Finally, *Well-being* refers to the more immediate state of equilibrium of the individual and his social–emotional environment. Klein postulates that most individuals suffer from some form of ill-being at various points in their lives. This may be, for example, in mass stress situations such as air-raids or geographical disasters, or on the other hand be the result of individual crises such as the death of a relative or difficulties at work. This state of ill-being is normally time-limited, with a return to equilibrium dependent on the nature and severity of the stress or the availability of treatment or other emotional assistance. If, however, equilibrium is not regained within a reasonable period, there may be a serious decrement in soundness or long-term health.

To return to Jahoda's illustration of physical health, a cold may be a temporary disruption of normal functioning; but an environment that causes a series of colds of sufficient severity can lead to an illness that results in pneumonia or permanent organic damage to the individual. Short-term signs of mental ill-health can similarly have long-term effects on the individual and the social organization in which he functions. 'Well-being' is thus of importance in itself, and there is a need, as Klein has suggested, to study acute states of 'ill-being' in their own right (irrespective of the degree of more central personality dysfunction) in order to learn about their aetiology in the individual, their immediate consequences in an organization, and the possible long-term effects on levels of sickness.

Other writers such as Eastwood (1971) and Wynne (1975) have also emphasized the considerable uncertainty that exists as to the significance of apparently transitory stress symptoms in people of previously 'sound' or healthy psychological constitution, and have pointed out that the precise relationship with more severe emotional illness is still unclear. Eastwood has stated: 'Information on who develops minor psychiatric illness and the natural history of these conditions is limited. . . . It is uncertain whether these individuals ultimately develop clinical states. . . . Practically nothing is known about the prevention of neurotic illness'.

The measurement of stress-related symptoms and states

Both authors of this chapter have been engaged (although in different organizations) with determining the amount and degree of psychological ill-health in working populations. However, the aims of our respective researches differed as to the aspects of mental health to be covered, and in consequence different instruments were used.

Stressors in the environment, which may be physical or psychosocial, vary in intensity, degree, and duration. The reaction in the individual will depend not only on these factors but on his susceptibility or vulnerability to stress, which is based upon the complex factors of his personality development, or in Klein's terms his soundness and stability.

Margolis and Kroes (1973) have suggested that there are at least five dimensions of job-related stress reactions that can be measured in order to obtain a total picture of the effects of job stress upon the worker. These are:

1. Transient short-term subjective reactions occurring in close temporal proximity to specific job stresses. Examples of these may be feelings of anger, fear, tension, and anxiety.
2. More chronic psychological responses that have become part of the individual's health status, rather than a reaction to specific work situations or events. These can be general malaise, constant fatigue, chronic depression or alienation.
3. Transient clinical–physiological changes such as alterations in levels of catacholamines, blood lipids, blood pressure, gut motility, etc., which indicate in an objective way that the individual is under stress. These changes may also be the precursors of psychosomatic or physical illness.
4. Symptoms of physical and/or psychosomatic illness such as coronary heart disease, gastro-intestinal disorders, etc.
5. Decreasing work performance. Increased error rates and increased incidence of accidents.

These five patterns of response may form the approach by which stress responses can be measured in the individual or the group.

It must, however, be recognized that there are problems in measuring all of these responses in some industrial situations. It is also desirable to bear in mind that all five categories of reactions may to varying degrees be influenced by personality traits, psychological conditions pre-dating the individual's entry into the workplace, or external factors unrelated to the work situation. The instruments to be used will depend upon whether the intention is to establish the number of individuals suffering from any kind of psychiatric disorder or to determine only those individuals *currently* experiencing temporary states of stress or ill-being irrespective of the soundness or stability of their psychological constitutions.

For instance, in a study carried out by the Health and Safety Executive, various questionnaires of the Symptom Check List type were used on a number of subjects working at employment rehabilitation centres. The most useful questionnaire in this situation was found to be the Middlesex Hospital Questionnaire (Crown and Crisp, 1966), as this accurately identified the more chronic reactions (dimension 2 above). This questionnaire gives a symptom profile by the use of six eight-item subscales—free-floating anxiety, phobias, obsessional, somatic, depressive, and hysteric. The MHQ, however, mixes long-term traits and symptoms, and therefore identifies both individuals with chronic personality difficulties and those with currently experienced symptoms.

In the research study carried out in the steel industry by one of the authors of this chapter, the term Psychological Well-Being (or Ill-Being) was adopted from the Klein model to indicate the type of health/ill-health aspect which was the area of initial concern. This is not to say that the long-term aspects were not considered important, for they were. But soundness and stability, to use Klein's terms, the long-term aspects of the mental health state, are normally already established when an individual enters industry, the results of an interaction between hereditary and early childhood factors. Our concern with soundness and stability must primarily be in appropriate placement and the better matching of personalities with tasks. Soundness and stability are not easily changed except by psychotherapeutic methods that are lengthy, not freely available, and not suitable for all individuals.

Concentration on the current state of equilibrium between the individual and his environment may, however, help us to learn how to influence the long-term aspect.

It was therefore our intention to measure temporary states of ill-being in those both of normally good and less good psychological constitution. Which methods therefore, of those available, were best for this purpose?

The measurement of transitory states of ill-being

In order to detect individuals displaying unhealthy levels of ill-being the simplest technique again is to use a questionnaire. The use of questionnaires and inventories for surveys of this kind (as opposed to individual clinical examination) obviously has many advantages of the grounds of feasibility, economy, ease of comparison, and objectivity.

In using such questionnaires to measure 'poor well-being', however, a number of points have to be considered. As already indicated, some previous surveys of poor mental health have attempted to measure a mixture of long- and short-term states, and this mixing of symptoms and traits has also been evident in the questionnaires used to determine levels of mental health and illness. Some of the possible drawbacks involved in various instruments are listed below:

(a) The tendency of certain questionnaires to measure the scatter rather than the intensity of symptoms. (An individual's effectiveness may be more impaired by one severe symptom than by several mild ones.)

(b) The inclusion of items that may confuse pre-morbid personality and early signs of illness.

(c) The tendency of certain personality types to 'choose' which symptoms they report. Thus questionnaires relying too heavily in the direction of either somatic or psychic symptoms might exclude certain kinds of *felt* psychological distress.

(d) Some questionnaires measure 'neuroticism' (a potentiality to have symptoms) that is more akin to long-term personality traits than to *present* felt distress or current ill-being.

(e) Some symptoms taken as indicators of emotional illness may be the transitory results of environmental pressures that will disappear if current crises are satisfactorily resolved. Such signs may therefore be healthy coping responses necessary to deal with an actual, and correctly perceived, external problem. This is not to say that such transitory stress levels should not be measured, but they should be measured only at an unhealthy level.

To detect individuals experiencing *current* psychological ill-being it is therefore necessary to use a questionnaire of a rather unusual type. In addition, problems associated with time require that the instrument be relatively short, acceptable to the respondents, and objective in the sense that the personnel administering it would not have to make any subjective assessments about the respondent. One questionnaire, however, matches these requirements fairly well. This is the General Health Questionnaire (30-item version) developed by D. P. Goldberg (1972) to identify individuals with minor emotional illness by assessing the severity of their emotional disturbance.

It was decided to use this instrument in the steel industry research described below; the reasons for the final choice of the GHQ are as follows:

— It does not confuse long- and short-term states.
— It does not place a heavy reliance on physical symptoms.
— Personality traits are not included.
— It does measure *current felt distress* in all individuals irrespective of their long-term personality condition.

— It excludes individuals with personality difficulties in a good phase.
— It does measure transitory symptoms but it is possible to determine different levels of severity.

Goldberg himself describes the emotional illnesses which the GHQ detects as mainly affective neuroses, that is, minor depressions, anxiety states, and anxiety reactions. A particularly useful feature of the questionnaire is that it detects people whose otherwise inexplicable somatic symptoms are accompanied by an affective disturbance which they have not presented to the physician.

A stress survey in the steel industry

Using the GHQ (together with other measures), a survey was carried out to determine variations in psychological well-being at a works undergoing reorganization, where the future of certain departments was in doubt. This reorganization entailed possible redundancies, although the numbers and timing were uncertain. Such a situation represents a crisis point in the lives of many people. As this was an exploratory study, the research had a number of aims. These were:

1. To examine the impact of a change or 'crisis' situation on the emotional health of the workforce. In this situation the 'crisis' was the possibility of a partial closedown, but it could equally well have been any other large 'change' event that altered past coping patterns. It was hoped that the information gained would help in counselling and other preparatory procedures in future similar situations.
2. To determine if general prevalence figures for emotional ill-health were reflected in a steel industry workforce.
3. To explore the best techniques of measuring 'stress' symptoms in the industrial situation.
4. To establish the relationship between high levels of psychological stress and absences for all forms of illness.
5. To identify stressful task areas and future research directions.

In order to obtain data for the above purposes the research was carried out in four stages. These were as follows.
1. The administration of a brief psychological measure (the GHQ) to 1300 operatives as part of a general medical survey. Half the questionnaire respondents came from departments whose future was uncertain and where redundancies might occur and half were from departments whose future was believed to be secure. As far as was possible, a similar balance of age and length of service groups was obtained for the different departmental

categories. An equivalent cross-section of jobs was included in each sample. Additional information was extracted from absence, sickness and personnel records.

2. A cross-section of those who completed the GHQ also attended a much longer interview. These interviews were based on a semi-structured questionnaire containing both open and closed questions. The questions covered the interviewee's attitudes and feelings to past and present work; his life experiences and views on current events; his interests and health. The interviews were to provide confirmation that people with emotional difficulties and high levels of stress were being correctly identified by the GHQ.

3. One hundred workers with minor stress conditions (as identified by the GHQ) were compared with a control group of individuals with low GHQ scores (non-stressed), matched for job, department, age, and length of service in order to assess differences in working time lost from absenteeism, lateness, early finishing, and sickness. This subproject was necessary to refine broader comparisons of sickness obtained for overall departments and groups in the first phase.

4. Finally, an in-depth and detailed case study was carried out on 20% of those studied in phase 3. Twenty-two individuals from the stressed group and 22 from the non-stressed group attended a $1\frac{1}{2}$–2 hr interview during which the GHQ was administered again. This second completion of the GHQ took place at least a year after the first administration.

During the course of the research, 1300 operatives had answered the GHQ and of these, at two different periods of time, 72 had attended subsequent lengthy interviews, 47 being from the group identified as stressed by the GHQ and 25 from the non-stressed group.

As indicated above, long interviews were held with a cross-section of positive and negative GHQ respondents at different score levels (or degrees of severity). These interviews were held in two stages, one set of interviews taking place shortly after the original administration of the brief psychological measure (the GHQ), the other being held just over a year later.

The first set of such interviews was carried out for the following reasons:

1. To ensure that people with emotional difficulties were being correctly identified. Particular attention was paid to individuals whose scores were unusually high, indicating considerable severity of distress, and those whose score was only just above the cutting line.

2. To ascertain the part that either the closedown uncertainty or work stresses had played in causing the 'score', as opposed to personal or home problems.

3. To determine interviewees' reactions to the completion of personality questionnaires of varying and increasing length, i.e. the 12-item MPI, the NSQ, and the 16PF. Form C (for future research purposes).

4. To determine the number of psychosomatic and minor physical symptoms experienced by GHQ respondents at different score levels.

5. To understand in more depth how an interviewee's personality had interacted with past and present stresses (within and outside of work) and ascertain how external interests may have aided endurance of stress. Such understanding has obvious relevance to the better identification of individuals most suited to isolated, monotonous or stressful tasks.

The second set of interviews, held just over a year after the first set, had the following aims:

(a) To give additional confirmation on the correctness or otherwise of the GHQ identification of stress levels (both in the first stage and currently).

(b) To determine the extent to which temporary stress symptoms appeared to have become chronic or to have triggered more serious emotional disorders, and in order to understand both the process by which this had occurred and possible relationships to long-term personality problems (preexistent and potential). Kedward (1969) suggests that the examination of the subsequent history of new cases indicates that, when symptoms have lasted a year (however mild they may be), this is also predictive of the outcome in three years. Those who have recovered in one year tend to remain free of symptoms and those who have not recovered tend to become chronic.

At the beginning of this research it was originally hoped to carry out a further set of interviews after three years, in order to confirm the proportion of those with no previous long-term personality difficulties whose high levels of temporary stress had escalated to more chronic forms of emotional disturbance.

Unfortunately, organizational changes in the steel industry prevented this further stage. Despite this, a considerable amount of useful information was obtained, both in the form of broad statistical comparisons and in that of individual in-depth personal case studies.

An outline of these findings is given below.

The findings

1. In non-closedown departments at this works, 5.1% of GHQ respondents had potentially unhealthy stress levels: in departments with a doubtful future, 10.6% of respondents reached this level. These figures indicate that while in normal circumstances the levels of stress in this steel industry workforce are certainly no greater than in a general population of similar age structure, a sharp and significant increase occurs during a situation of change and uncertainty.

2. Although the percentage of those who exceeded the GHQ cutting score was much higher in potential closedown departments than in other

departments, in the sample of GHQ respondents interviewed the closedown stress was (in most cases) only one of a number of stresses in the interviewees' lives. However, in view of the considerable difference in the total of stressed cases between departmental categories (closedown/non-closedown), it seems clear that while the closedown possibility was a contributing rather than the only stress, in many individuals it was sufficient to increase ill-being to a level likely to be impairing.

3. In this specific situation the overall percentage of the shop-floor population with unhealthy stress levels, as indicated by the GHQ, was 7.55%. To this can be added a further 2.5% identified from other sources (not through the GHQ survey), thus increasing the total to 10%. Even this is by no means the full total, as the survey measured only present distress, not long-term conditions in a good phase, nor did it include all physical conditions with associated emotional factors.

4. A relatively small portion of the works population experienced a disproportionate amount of sickness of all kinds, emotional and physical. For instance:

(a)   Those who exceeded the GHQ cutting score had more sickness and unofficial absence than the general population (32 days against the general average of 19). When stressed individuals were compared with a non-stressed control group matched for age, job, and experience, the difference widened even further (40 days against 20).

(b)   Those patients with a stress or psychological diagnosis on the medical file (5.7% of the total population) accounted for 22.8% of all days lost for all types of illness (physical and emotional).

5. The information obtained has shown the GHQ to be a valid and acceptable indicator of stressed individuals in the industrial situation. This was confirmed in three ways:

(a)   By the higher levels of positive GHQ scorers. People who exceeded the cutting score on the GHQ as a group had more illness absence than those who did not. This was true of the research period and over the previous seven years.

(b)   By extended follow-up interviews with a cross-section of high and low GHQ scorers aimed at determining both long-term personality traits and recent life crises within and outside work.

(c)   By the number of psychosomatic symptoms reported by GHQ positive scorers compared with negative scorers.

6. The results of the survey and subsequent interviews also indicated certain task areas with higher than average levels of stress and which pointed

to a need for further investigation, in particular crane driving, labouring, and certain isolated or monotonous jobs. The interviews also emphasize the importance of monitoring the long-term effects on personality over time, as well as the more immediate effects from current stresses. Responses and life histories obtained in the extended follow-up interviews suggest that some forms of adaptation to reduce short-term stress may have been at the expense of long-term flexibility, psychological health, and future flexibility.

Discussion

It is very obvious from this and earlier studies that stress-related symptoms and disorders are so extensively distributed in any workforce that it is desirable to put the total problem in perspective. This can most easily be done by referring to one of the social psychiatric surveys mentioned earlier, that of Taylor and Chave, carried out in the sixties on the population of a small English town. The results of this specific investigation are of particular interest in that they enable the steel works findings to be set in a wider context, and also help to confirm the percentage of symptoms that may be triggered by *current* stresses.

The implications of Taylor and Chave's findings can be summarized in the following way:

1. About 70% of the population can be described as having good psychological health. This does not necessarily imply that people falling in this category never suffer from symptoms of stress, but such symptoms are only experienced very occasionally and do not tend to reoccur.

2. Approximately 25% of the population are born with, or develop in early life, nervous systems that are particularly prone to minor psychological and psychosomatic illness under relatively moderate stress. The majority of these people are not, however, under medical care, although they tend periodically to experience minor symptoms of various kinds and in ways which may affect their well-being or efficiency. This group has been defined as the subclinical syndrome group. The total percentage at risk remains fairly constant in size.

3. There is another group of people, who, like the subclinical category, are also constitutionally anxious and tense but who, because of long-term personality traits, tend to suffer from much more severe conditions that cause them to experience ill-health all or most of the time. These people probably worry even when most obvious external pressures are removed. This group represents 5% of the total population.

4. Finally, there is a group of uncertain and varying size of basically healthy people who, although they have not previously exhibited symptoms of poor psychological health, do under certain forms of severe social or environmental stress also display symptoms very similar to emotionally ill

patients under treatment in a clinic. This group is described as suffering from 'environmental neurosis'. Their reactions are assumed by these researchers to be transitory, but other research such as that of Leighton *et al.* (1963) suggests that once stress symptoms have been triggered to an impairing level they tend to become chronic and reappear more easily in future. The potentiality for illness, both psychological and physical, may therefore become much greater.

What percentage of all those at risk in the four categories described escalate to more severe states, or come to the attention of medical agencies under the influence of current stress?

The picture suggested by this and some other surveys in the literature (if allowance is made for differences in terminology) would seem to be this: the 5% of the population with the more severe constitutional neurotic conditions (because of their susceptibility to very slight stress) suffer constant minor illness and visit their doctors most of the time.

This chronic 5% tends at periods of crisis to be added to in varying proportions both from the subclinical syndrome population and from those who were previously healthy. Although the numbers of those in the two later groups whose symptoms have escalated can vary, together they tend to total another 5–7%. Thus the chronic 5% is almost constantly increased by environmental or social pressures to a total of 10% with severe or impairing symptoms. This variable 5% has been confirmed by the steel works study.

General practice studies, as shown by Kedward (1969), suggest that a small proportion of new cases become chronic and that this is sufficient to replace the equally small proportion of chronic cases that get better each year. Thus the chronic 5% is constantly maintained by new cases triggered through current stress. Prevention of the additional 5% is at least theoretically possible by appropriately timed intervention or other means. This may seem a small proportion of the total problem, but it is the chronic 5% that suffers from a disproportionately large percentage of sickness of all kinds, physical and emotional. This also has been confirmed by the steel works study.

If therefore we can prevent a significant proportion of new recruits being added to the emotionally ill population, this could have a disproportionately large effect in reducing the sum of days lost for illness absence in industry.

Such prevention, however, entails being able to identify and prevent unhealthy levels of stress and ill-being at a preclinical stage when intervention is easiest and most economic of resources.

The prevention of emotional ill-health in the workplace

It is possible, as has been shown above, to measure and confirm the proportion of those suffering from temporary states of psychological ill-being in an industrial workforce, and also to prove how this is related to higher levels of actual sickness.

To what extent, however, is it possible to prevent such high levels of transitory stress escalating to more chronic conditions?

It is certainly true that the long-term aspects of mental health (the more enduring personality traits that increase vulnerability to stress) cannot easily be changed except by lengthy psychotherapeutic methods, and these are neither easily available nor suitable for all individuals.

The body of thought known as 'crisis theory' (Caplan and Grumebaum, 1964) suggests, however, an alternative approach more relevant to problems of psychological well-being in the working environment.

This approach places the main emphasis on improving the ability to cope with current events and ignoring or putting less emphasis on past personality difficulties. The crisis theorists suggest that stress symptoms are displayed in those both of sound and less sound psychological constitution when past coping abilities are no longer adequate for a new situation or threat. The person concerned may be unable to deal effectively with a newly arising problem either because his own abilities or resources are insufficient or because his social or organizational environment has not provided him with the means to cope adequately with the new situation.

Crises can vary in severity from those large-scale events that will have an impact on most people, irrespective of their previous psychological stability and health, to minor crises that may affect only particular individuals whose psychological constitution is such that they have a narrow range of coping capacity.

What is common to both categories is that the crisis exists because past techniques for managing problems are exceeded. If the individual evolves, or is helped to evolve, new methods to cope with the situation within a reasonable period, this will result in psychological growth, and the experience of successful coping will make him that little more confident in dealing with future problems or situations of uncertainty.

All crisis events, however, represent points of vulnerability to future breakdown. This is because in all circumstances where a person is unable to cope, memories of past failures or personality difficulties tend to be reactivated. If the individual fails again in this new situation of difficulty, the old problems are reinforced, he becomes more rigid and less adaptable, and has even less confidence in dealing with future unfamiliar situations. His potentiality is thus decreased, and he will tend to avoid any circumstances which present similar threats or unpleasant feelings of excessive stress.

Neurosis has in fact been characterized (by Angyal, 1965) as a state where there is an overemphasis on security, where an individual devotes his entire life to protecting himself from threats. In such circumstances all decisions are overwhelmingly centred on dangers to be avoided, rather than objectives to be achieved. As a consequence, all coping responses become more rigid and the individual is likely to feel stress in a wider range of situations, and to react to such stress with psychological or physical illness, dependent upon his personality.

Nevertheless, it needs to be emphasized that all crisis situations tend to be occasions where the individual is particularly open to influence for either good or ill, and are thus periods when the balance may be tipped towards either a more or less effective repertoire of coping responses, and towards better or worse emotional health.

It is important, however, that help to resolve such situations is appropriately timed, that is, during the stage of crisis before rigid defences or maladaptive and neurotic solutions have become consolidated in the personality. This concept has obvious relevance to those situations where people are faced with change in their working life, whether this involves technical changes, job alterations, or the restructuring of organizations. It also emphasizes the need for the appropriate timing of preparation and assistance to those in change situations.

Although the relationship between unsuccessful crisis resolution and emotional illness has not been entirely proven in populations as opposed to individuals, the steel industry survey provides confirmation of the effect a crisis situation has on both short- and long-term emotional conditions. Crisis theory also fits in with other findings in stress research.

Psychiatrists of the University Medical School at Rochester, New York (Engel 1968), for instance, describe a psychological state that follows stress and results from the failure of adaptive resources. This, they believe, constitutes a common setting for the onset of illness, both organic and psychiatric. They call this the 'giving-up complex', a description taken from the characteristic remarks made by people in this condition.

The individual in this state, they suggest, may, dependent on his personality, follow four directions:

1.   He may regain the ability to cope through his own efforts, or with external help if this is appropriately timed.

2.   He may become physically ill.

3.   He may develop a psychological illness.

4.   He may adopt socially deviant behaviour, or act out his symptoms by absenteeism, aggression, lateness, or accidents.

Hinkle and Wolff (1958) and Holmes and Rahe (1967) have also shown that ill-health, whether predominantly physical or psychological, occurs in clusters, usually following a series of life changes, or coincides with the times when an individual is experiencing difficulties in coming to terms with his life pattern. Hinkle and Wolff inferred from their research that all individuals are particularly prone to illness when they perceive their life circumstances to be threatening, although the same circumstances may not necessarily appear so to the bystander.

Foulds (1976) has in fact stated that 'it is the threat of a catastrophic lowering of self-esteem which triggers *all* non-organic psychiatric illnesses— and the various forms of illness are in part the resultant of different modes of defence against this threat'.

It is clear from what has been said that temporary states of ill-being springing from crisis points and stress are important areas for our initial concern, both because of their immediate consequences and because of their possible long-term influence on more chronic conditions, or their limiting effects on the full use of each individual's potentiality.

It is, however, necessary to emphasize again that acute transitional or 'crisis' points in life are not only threats to well-being but also opportunities for better mental health and psychological growth. In industry, technological, organizational and job changes all provide situations where well-being may be affected, and individuals may move nearer to, or further away from healthy adaptive patterns of functioning.

If individuals in change situations or at points of personal crisis are able to extend past techniques or experience new coping abilities, this can result in greater flexibility in future and an increased ability to withstand future stresses; the potentiality to illness, psychological or physical, may therefore also be decreased. People who are helped to cope in these situations may also, through the experience of successful coping, begin to resolve old personality problems without the necessity for lengthy therapy.

Staff involved in helping, preparing and training people to cope with new situations, new jobs or stressful events are in a significant position to improve mental health, as all people are most open to influence at transitional points. Intervention to prevent maladaptive reactions is, however, most effective (and easiest) when timed near to a 'crisis' and before an individual is defined by himself or others as ill. This is of particular importance in view of the limited medical resources in the area of emotional and stress disorders.

Interviews with a cross-section of high scorers on the GHQ also underlined a need for counselling services to be extended, or more specifically directed at emotional ill-health in the workplace. This would enable individuals to be helped *before* their symptoms escalated to illness behaviour, rather than after. It was very evident that some of those interviewed would have sought such help if it had been available. The need for such a work-related service was also confirmed by a 'Samaritan' operating in the area.

Counselling does already exist in various forms throughout industry and valuable help is given to people with problems through these channels. Many people in the nursing, welfare and personnel areas have experience and interest in problems of stress, and already intuitively manage covert forms of psychological difficulty; many of these would welcome the opportunity to gain and help build up additional specialized knowledge in this area.

Effective lay counsellors able to deal with certain aspects of this problem could also possibly be selected and trained from indigenous resources within the workplace, provided, that is, they are supervised and backed with the necessary supporting expertise. Such indigenous counsellors who know the work environment have advantages over other agencies without such knowledge.

In addition, there is a need for much greater awareness by management of the significance of 'stress' symptoms in others and themselves, in order that self-help or intervention can occur before minor conditions escalate to more serious conditions requiring medical attention. A great deal of minor emotional disturbance in industry is more appropriately dealt with by social intervention than by the use of scarce medical facilities. Once symptoms have become 'illness', the time and skill needed to deal with them become much greater.

One final point needs to be made. The emphasis in this chapter has been on the importance of identifying stress symptoms at an early stage before maladaptive reactions become an integral part of the personality. This is a logical approach both on the grounds of preventing future illness and because crisis points (minor or major) provide an opportunity when people can not only be helped with their current difficulties but may also, because of their 'openness' to assistance at such times, be helped to at least begin the resolution of preexistent personality problems and maladaptive defences.

The Steel Survey for these reasons concentrated on current distress. It was very clear, however, from the intensive follow-up interviews with both high and low GHQ scorers, that some individuals with currently low stress scores had 'adapted' to certain tasks in ways that had diminished stress at the cost of long-term mental health.

There is a need therefore to consider both the short- and long-term effects of tasks on mental health. For instance, there are situations where an individual with a high level of 'soundness' (good long-term mental health) may, after some time in a stressful job, begin to show symptoms of ill-being (short-term mental health). This can either be the result of excessive demands on coping abilities or too little demand, depending on the individual's personality. In the case of monotony, isolation, and under-demand, the 'sound' person may 'adapt' at the expense of his soundness; although his surface symptoms may decrease, his future potential may be less.

On the other hand, some individuals who choose jobs where coping demands in certain directions are less may be cut off from circumstances where their coping responses could be widened.

Our research underlined the additional knowledge that has to be gained to decrease the level and effects of psychological ill-being in the workplace. Much more needs to be known about:

(a)   The long-term effects of unhealthy forms of short-term adaptation.
(b)   The trade-off that appears to occur between potentiality and well-being (i.e. the apparent sacrifice of self-actualization in order to preserve current well-being).
(c)   The interrelationship between well-being in work and private life.
(d)   The personality characteristics that tend to increase or prevent psychological stress in certain task areas.

The need for such further research in this area is of particular importance for, as McKeown wrote in 1961, 'emotional disorders will constitute the major health problem over the next forty years.'

References

Angyal, A. (1965). *Neurosis and treatment. A Holistic Theory*. John Wiley, New York.

Caplan, G., and Grumebaum, H. (1964). *Perspectives on Primary Prevention*. Basic Books, New York.

Collins, R. T. (1961). *A Manual of Neurology and Psychiatry*. Grune and Stratton, New York.

Crown, S., and Crisp, A. H. (1966). A short clinical diagnostic self rating scale for psycho-neurotic patients. *British Journal of Psychiatry*, **112**, 917.

Dohrenwend, B. P., and Dohrenwend, B. S. (1965). The problem of validity in field studies of psychological disorder. *Journal of Abnormal Psychiatry*, **70**, No. 1, 52–69.

Eastwood, M. R. (1971). Screening for psychiatric disorder. *Psychological Medicine*, **1**, 197–208.

Engel, G. L. (1968). A life setting conducive to illness. *Annals of Internal Medicine*, **69**, 293.

Ferguson, C. A., *et al*. (1965). *The Legacy of Neglect*. Industrial Mental Health Association, Fort Worth, Texas.

Foulds, G. A. (1976). *The Hierarchical Nature of Personal Illness*. Academic Press.

Fraser, R. (1947). *The Incidence of Neurosis among Factory Workers*. HMSO, London.

Goldberg, D. P. (1972). *The Detection of Psychiatric Illness by Questionnaire*. Institute of Psychiatry, Maudsley Monographs, Oxford University Press.

Harrington, J. A. (1962). Research into neurosis in industry. In *Aspects of Psychiatric Research*, Oxford University Press.

Hinkle, L. E., and Wolff, H. G. (1958). Ecological investigations of the relationship between illness, life experiences and social environment. *Ann. Intera. Med*, **49**, 1373–88.

Holmes, T. H., and Rahe, R. H. (1967). The social readjustment rating scale. *Psychosomatic Research*, **11,** 23.

Jahoda, M. (1958). *Current Concepts of Positive Mental Health*. Basic Books, New York.

Jones, H. G. (1966). Neurosis. In *Progress in Mental Health*. Office of Health Economics.

Kedward, H. (1969). The outcome of neurotic illness in the community. *Social Psychiatry*, **4,** No. 1.

Klein, D. C. (1960). Some concepts concerning the mental health of the individual. *Journal of Cons. Psychology*, **24,** No. 4, 228–93.

Leighton, D. C., Harding, J. S., Macklin, D. B., Macmillan, A. M., and Leighton, A. H. (1963). *The Character of Danger*. Basic Books, New York.

MacIvor, J. (1964). Industrialisation and Mental Health. Paper read at World Federation of Mental Health.

Margolis, B. K., and Kroes, W. H. (1973). Occupational stress and strain. *Occupational Mental Health*, **2,** No. 4.

Markowe, M., and Barber, L. E. (1953). Psychological handicap in relation to productivity and occupational adjustment. *British Journal of Industrial Medicine*, **10,** 125.

National Association of Mental Health (1971). *Stress at Work*. Mind Report No. 3.

Neff, W. S. (1968). *Work and Human Behaviour*. Atherton Press, New York.

Srole, L., Langner, T. S., Michael, S. T., Opler, M. K., and Rennie, T. A. C. (1962). *Mental Health in the Metropolis*. Mid-Town Manhattan Study, Vol. 1. McGraw-Hill, New York.

Taylor, Lord and Chave, S. (1964). *Mental Health and Environment*. Longmans, London.

Wynne, L. C. (1975). *The Comprehensive Text Book of Psychiatry*, Chapter 25. The Williams and Wilkins Company, Baltimore.

PART III

Designing the Stress out of Work: Ergonomics and Job Design

Introduction

Both sides of industry are beginning to take a greater interest in working conditions. The problems of industrial health are becoming more and more publicized. These trends can be explained partly by economic and technological advances in the industrialized world. Relative economic prosperity has diverted the attention of the working population away from simple monetary rewards to compensate for unacceptable working environments. Health, living conditions, transport, and leisure amenities all contribute to the standard of living of a modern industrial society. Individual workers are refusing to accept unsatisfactory working conditions and this is being demonstrated openly through union policy or covertly by increased absenteeism and turnover. In many European countries unions are taking working conditions as one of their priority areas for action. This is especially true for certain boom industries where employees are paid high wages. In the North Sea oil industry the unions have traditionally had a very low membership. However, after a number of serious accidents on oil rigs, safety has become an important issue on which the unions can have some influence. Technological advances in the nuclear and chemical industries have also had their influence in changing the attitudes towards working conditions. The potential dangers of environmental pollution from ever larger installations has made surrounding populations conscious of the need for controls. Industrial health problems are no longer confined to the factory but can affect the surrounding areas as well.

The case for improving working conditions in industry can be justified in a number of ways. It is only natural that increased standards of living in industrialized countries will be accompanied by a desire to improve the overall quality of life, which must include the quality of working life. Legislation is gradually being introduced in a number of Western European and Scandinavian countries to provide for increased worker representation in the decision-making process within the firm, for wider controls on safety and hygiene, and for the introduction of education and training programmes for employees. This social legislation is not being introduced solely on humanitarian grounds. It has been realized that the consequence of unacceptable working conditions can be costly to society as a whole through loss of production and increases in social security costs. Thus industry has a responsibility to society to provide working conditions which fit the worker as well as fulfilling production criteria.

The interest shown by industry in improving working conditions has often been hindered by the lack of techniques or methods necessary for the implementation of results of applied research in the work sciences. To close the gulf between applied research carried out by academics and application to industrial problems still needs considerable effort. One of the aims of the PROMSTRA group is to provide a link between the academic approaches to industrial problems and the application of research data in industry. The results from fundamental research are often not sufficient nor presented in a manner which is useable in applied problems. Similarly, it is not always certain whether the questions which are being studied are relevant to the problems found in industry. Industry, both management and unions, should be encouraged to collaborate with research workers in research on working conditions, to help direct the research to relevant problems, and to provide facilities where it can be carried out. The PROMSTRA seminars have provided a forum for discussion between representatives from industry and research groups, one of the intentions being to increase the amount of collaborative studies being pursued.

Parts I and II have presented chapters which describe methods of assessing physiological and mental workload and occupational stress. The difficulties met when applying these methods in industry were also discussed. PROMSTRA does not intend to remain only at the analysis stage when studying working conditions, but has the additional objective of promoting practical ways of improving jobs to avoid overloading the worker or damaging either his physical or mental health.

The chapters in Part III describe some techniques used in job design, give some advice on designing jobs which are better fitted to the worker, and lastly give examples of jobs which have been improved by the use of ergonomics.

The first two chapters describe techniques which have been used in industry to design tasks. The first of these introduces a matrix method of analysing the different variables associated with a repetitive routine task. The author describes how the results from the analysis can be used to improve job content. Dr. Aberg has frequently used this technique in Sweden, notably in the steel industry. Dr. Bosman presents a systems approach to designing production systems. The method emphasizes the importance of considering the human operator and his requirements in system planning and building. Again this is a practical method, which Dr. Bosman has used in the aircraft industry.

The two case studies reported next deal with two very different working environments, one a commercial administration of a large electronics firm and the other a press workshop in a light engineering company. The introduction of information-processing systems in industrial organizations presents numerous technical problems. These cannot be dissociated and viewed independently of the jobs done by the people working within the system. The mismatch between user requirements, in terms of task design, information presentation or communication, and the systems analysis perception of these requirements occurs as a result of trying to consider the technical aspects in isolation. The example given in this chapter involves the implementation of an on-line ordering system for electronics components in a large commercial department. The design of the information-processing system had direct consequences on the job content of the clerks, who had previously used a classical filing system. The repetitive form-filling and filing tasks were changed, but they had not always been replaced with tasks which fulfilled the necessary design criteria. The employees who remained in the ordering department were required to use the new system and the modifications to their jobs often resulted in them having to carry out a lower-skilled job. The introduction of VDU terminals meant that all levels of employee or manager who wished to consult the system needed to have basic keying skills. Although this may have been a temporary problem, as keying skills will become more common with the introduction of VDU terminals in all types of work, it was quite overlooked. In conclusion it appears from this study that it is essential that those expected to use the new system should participate in the planning and implementation of the system if it is to be accepted and used effectively.

The second case study was commissioned by a government agency, and was intended to be an example of the contribution an ergonomics study can make to improving working conditions. The light engineering firm where the study was carried out had planned to move to new factory premises and wished to improve the working conditions. The aim of the study was to provide suitable recommendations after analysis of the working conditions in the old factory unit. The chapter attempts to present the methodology the research team used to carry out the ergonomics analysis. It emphasizes a global approach to the study of working conditions, which is to say that the analysis is based on a study of the work actually done by the operators but must not neglect the numerous other factors in the social and economic environment which influence the work on the shop floor. A preliminary study of the economic, social and organizational factors within the firm was thus necessary in order to define the hypotheses used for the analytical phase on the shop floor. This phase used mainly observational and interview methods for collecting data, consequently it was essential that the operators were willing to cooperate in the study if valid results were to be obtained. The research team found it necessary to explain the aims and methods of ergonomics to all sections and to all levels of the firm's hierarchy to avoid any misinterpretation of the study's objectives. It is interesting to note that the team was also obliged to explain the limitations of ergonomics recommendations, as they were asked to go further than the original project brief.

The last two chapters give points of view on the trends in working conditions. Leymann presents a personal view of the development of research in working conditions in Sweden since the war. Research in this area is often controversial, and Leymann remarks that as research workers become more interested in bringing their work to the shop floor they find that they do not have the same encouragement from the authorities as for university based studies. This is possibly because the objectives have changed from being primarily production orientated to being concerned with the health and well-being of the worker. Leymann also suggests certain new areas that research in working conditions will be concerned with in the future. He considers that the most important change will be an increase in democracy in the workplace, allowing workers a greater influence on decision-making in the firm, which will then affect the different issues considered worthy of research effort. Leymann suggests that these will include classical areas such as safety, toxicology and workplace design but will also cover work organization and research on the development of codetermination in the workplace.

Part III finishes with a chapter by the General Secretary of a French metallurgical union, presenting the union's general policy on certain issues in working conditions. The union considers working conditions on the shop floor as one of the priority areas for union negotiations. It has developed a training package for the use of the local representatives. The package aims to train the representatives to make an evaluation of their own working conditions and gives the workers, who are in effect the 'consumers', a means of criticizing the different types of working conditions they are subjected to. Leduc also gives a number of examples where the union has intervened and ensured that conditions have been changed for the benefit of the employees. The union's role concerning working conditions is in some ways ambiguous; Leduc makes the point that it is not to design jobs or workplaces but to protect the health and welfare of its members. The union's function is essentially to act as a control on the types of working environment its members are obliged to work in; consequently any statement on its part must necessarily remain very general.

As in Parts I and II, the chapters presented here cannot claim to have covered all the aspects of improving working conditions. However, the range of topics discussed gives an idea of the problems which are present in industry and demonstrates that both sides of industry are looking towards the field of work design for changes to be brought about. It has become apparent that the jobs found in industry do not fulfil the criteria for good job design and are increasingly not being accepted by those who have to carry them out. The costs in terms of damage to health and disatisfaction resulting in loss of production will no longer be tolerated. Changes in working conditions and progress in job design can only be made if all the groups in industry are suitably informed of the methods available for tackling occupational stress and improving the quality of working life. The PROMSTRA seminars and this collection of papers hope to go some way to help to disseminate this sort of information.

Stress, Work Design, and Productivity
Edited by E. N. Corlett and J. Richardson
© 1981 John Wiley & Sons Ltd

Chapter 10

Techniques in Redesigning Routine Work

Ulf Aberg
Royal Institute of Technology, Stockholm, Sweden

In the various attempts at improving the work environment, the physical factors such as noise, climatic conditions, air pollution, etc, are the most straightforward to deal with. In many ways, though by all means not all, they can be attacked independently from other aspects of job design. This is more true at the factory level than at the machine design level, and it is of interest to note that the more fundamentally you try to solve the design problems, the more you have to approach the problem from a many-sided point of view, that is from a true ergonomic angle.

If we consider mainly the physiological aspects of work and those which are at least partly related to the use of the sensory organs and the processing of information, stricted (though important) areas like chairs, instruments, problems of lifting, etc. But there are many experiments on a more ambitious scale, in England, France and Sweden. A common feature of these is that they are involved at a more basic level than just improving the immediate cause of strain. The ergonomic/technical solutions here involve climate, work load, dust, work positions, etc., but they should also centrally involve the psycho-social situation, in the sense that the freedom and the variability and the skill of the former work are retained. The heart or kernel of this kind of work is that it opens up possibilities for a solution possessing very many degrees of freedom.

However useful these attempts may appear for the workman concerned and no matter how intricate and interesting the engineering and other design problems may be, it seems that the solution processes have not achieved a degree of effectiveness sufficient to cope with the problem of routine work and perhaps especially that of the machine line.

Now why is this so? May it be that, after all, mankind has a soul which needs some nourishment from work which short-cycled, repetitive, non-responsible, non-skilled work cannot provide? May it also be that implementing the conventional philosophy of industrial job design has driven this kind of work into a corner, so that there is no obvious way out?

We are all very well aware that many believe that on one of the lowest shelves in hell you will find a man called F. W. Taylor. Well, let us leave him there for some time and look a little at the early design criteria for routine work and also, as a matter of interest, at several criteria of the same school in more recent times.

If you open a book on method studies, you will probably find examples of the objectives like these.

1. Increased profitability
2. Increased production
3. Decreased costs
4. Decreased labour requirements (in number)
5. Decreased spare capacity
6. Decreased accident risk
7. Decreased work load
8. Decreased need for skill

We note that production efficiency and economy are the main points here, but also that there are definitely some human objectives (work load and accidents). A closer investigation reveals that for the economic and production objectives you have routines to help you solve the problems and check the results, but for the human factors the control methods and the work methods are in most cases *ad hoc* and subjective.

Later on, ergonomists started to develop methods for the design and the redesign of work. Stephen Griew, (1964) for example, in a report on job redesign for older workers for OECD, stated explicitly that you have to know two things to be able to redesign a job for older workers: firstly it is essential to know how ageing affects working capacity and performance, and secondly one must know how to recognize in a job features which are likely to militate against the performance of the older worker.

These are certainly very sound ergonomic principles. They do, however, concentrate on physical characteristics and might also tend towards the same efficiency trend as the method study objectives.

Singleton (1967), in a paper on systems design, points out that 'all systems exist to serve human needs and must ultimately be directed by human decisions'. He adds, however, that the real systems design problem is not to allocate functions between man and machine but to delegate functions from man to machine.

This is not always true, and from a human point of view it might be better to delegate functions from machine to man. It is interesting to note that industry, partly out of necessity and partly because it consists of human beings open to opinions and ideas, has abandoned the single-minded efficiency concept and is now seriously considering the human aspects. True, these are mostly formulated in the following way:

> Decreasing labour turnover
> Decreasing absenteeism
> Etc.

These objectives or wishes express what we might call an equifinality, but they are not very operational in the sense that you can start designing a job with the help of them.

It is as if we still had the old situation, a series of restrictions and requirements for the human being, considerably more humanized than before, but still the same design philosophy. What one would like to do is to turn the wheel around and start the construction of the job from the human being.

Let us leave the question of equifinality and try to formulate the demands of production and the demands of the human operator.

The requirements, to be usable in practice, must conform to certain standards:

1. It should be possible to develop them in detail from some fairly general statements.
2. They must be measurable, so that you can test your design against them and see if it stands up to the measure.
3. It should be possible to select different levels of comfort.

It is obvious that criteria like 'increased job satisfaction', 'decreased labour turnover', etc., are not suitable according to these standards. Objectives from the viewpoint of the company could be better stated, for example, as:

What category of people are expected to work in the company (sex, age, handicap, etc.)?
What level of knowledge or skill do we wish?
What kind of development for the personnel?
What guidelines should be valid for the rotation of people within the company?
What kind of work organization is appropriate?

These criteria form a personnel policy, or part of it. The more detailed criteria can be considered as subsets of these. For example, the physiological criteria are dependent on whether you have all ages, women, etc., in the workforce (Table 10.1).

Table 10.1   Primary physiological criteria

1.   Work load must not mean health risks.
2.   Work should not be perceived as heavy by the selected personnel.
3.   Work should also be possible for non-selected personnel to perform.
4.   Work should not be perceived as heavy by non-selected personnel.

Routine work often gives rise both to dynamic physiological overload, caused by frequently repeated lifting motions, and static overload, mainly caused by awkward postures. Additional features of routine work are the psycho-social drawbacks like monotony, lack of communication with others, confinement, lack of work content, etc.

The psycho-social criteria for acceptable work are, at least to engineers, less well known than the physiological or environmental criteria. They are not very often used by technical people and it might be worth while presenting a set of them (Table 10.2).

Table 10.2   Psycho-social criteria dealing with cooperation

1.   The operator can influence working rate.
2.   The operator can vary the working method and has opportunities to maintain visual contact with others and can carry on conversation at close quarters.
3.   The operator can contact fellow-workers during work in order to deal with questions arising from the work task.
4.   The operator can continuously determine working methods and working rate in cooperation with others.

As can be seen, the objectives have been ranked in increasing degree of complexity. The idea is that suggestions for solution could be made at each level and discussed with reference to their technical feasibility, cost, time to achieve target, etc., after which a decision can more easily be taken about what practical level can be achieved. A second example can be given concerning rest pauses. Breaks in work are of course necessary in heavy physical work and in that connection are well known. In routine work, breaks are not mainly a physical problem but rather a means of relieving monotony and allowing a greater personal freedom in work. Rest criteria can be formulated as set out below (Table 10.3).

Table 10.3   Psycho-social criteria dealing with rest pauses

1.   The work pace can be influenced by the operator.
2.   The operator has the opportunity to create buffer stock.
3.   The operator has the opportunity to leave the workplace.
4.   The operator has the opportunity to plan his own work over a whole day or large part of a day.

These criteria are also ranked in increasing degree of complexity and are of the same general character as those for cooperation. In part they are even identical, which of course only means that they represent some fundamental conditions that must be fulfilled for improvements in both categories of requests.

The first phase of an ergonomic project is usually a measurement of the different stress factors of work, where the physical and chemical environment as well as the physiological and sensory loads play an important role. It is obvious from the foregoing discussion that psycho-social factors also should be measured or noted. In addition, it is important that not only the negative factors should be included in such a listing but also the positive side of work, as it is as necessary during a change that the positive

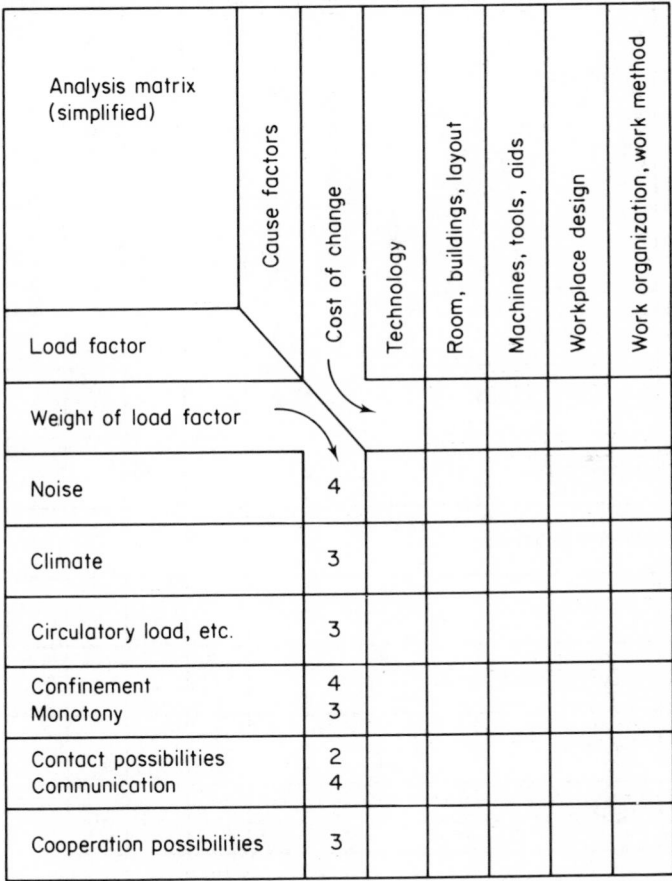

Figure 10.1   Analysis matrix used for mapping the relation between load factors and cause factors

features should be preserved as it is that the negative side should be eliminated.

To facilitate the later stages of the project work, it is of great practical value to list the origin of the load factors at the same time. This could be done in the form of a matrix, which gives a very simple and clear (though by no means exact) picture of the complex relationships in work (Figure 10.1). The matrix could with advantage be used in the practical design work if the design suggestions were put in place of the cause factors and their estimated positive or negative influence on the load factors noted. In this way it is easy to sum up the total effect on the work situation (Figure 10.2).

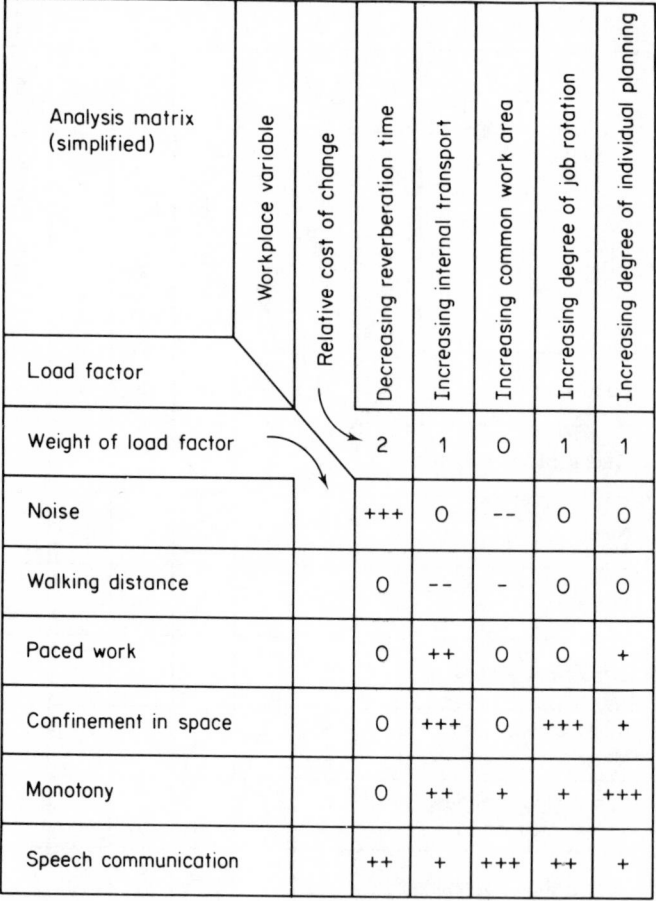

Analysis matrix (simplified) Load factor	Workplace variable Relative cost of change	Decreasing reverberation time	Increasing internal transport	Increasing common work area	Increasing degree of job rotation	Increasing degree of individual planning
Weight of load factor		2	1	0	1	1
Noise		+++	0	– –	0	0
Walking distance		0	– –	–	0	0
Paced work		0	++	0	0	+
Confinement in space		0	+++	0	+++	+
Monotony		0	++	+	+	+++
Speech communication		++	+	+++	++	+

Figure 10.2   Analysis matrix used for testing effects of changes in workplace variables on load factors

The discussion above has been concentrated on how to formulate objectives and on the technicalities of achieving certain goals. It will become increasingly important to involve workers and similar categories in job improvement and job redesign. Without going very deeply into this indeed very vast subject, it can be stated that participation will create a need for a very thorough dissection of the problems of work design. This chapter was intended to give an outline and some suggestions for such a procedure. A proper method is always a help in solving any kind of problem, but it must also be tacitly assumed that a good solution to an ergonomic problem always rests with the knowledge and imagination of the designer.

References

Griew, S. (1964). *Job Re-design.* OECD, Paris.
Singleton, W. T. (1967). The systems prototype and his design problems. *Ergonomics*, **10,** No. 2, 120–4.

Stress, Work Design, and Productivity
Edited by E. N. Corlett and J. Richardson
© 1981 John Wiley & Sons Ltd

Chapter 11

Systematic Design of Socio-technical Systems

D. Bosman
Twente University of Technology,
Enschede, Netherlands

Introduction

'A socio-technical system is a construction, composed of equipment, procedures, and rules/regulations, which, under the control and supervision of humans, must accomplish a given goal or mission.'

This definition embraces the trivial (such as the cyclist among traffic), through complicated systems (e.g. industries), to the very complex (such as societies and cities).

Artificial means (equipment, procedures) must be designed such that:

— Good and reliable use is made of physical laws;
— Due consideration is given to human and instrument capabilities.

This chapter is particularly concerned with the second requirement.

The human element in the system

Human factors, or ergonomics, is a (multidisciplinary) technology with the following objectives:

— To optimize the mental and physical functioning of a person in his or her work situation;
— To improve the quality and quantity of the product of his or her work.

To this end, knowledge and methods from the technical, organizational, behavioural and biological sciences must be introduced in the design of man–machine systems (MMS).

The need to apply ergonomics is increasingly acknowledged, one reason being that this technology is gradually coming of age, the other that information processing is no longer a vaguely understood faculty reserved solely to man. Also, it is recognized that many MMS provide very good service over a large part of their 'capability profile' because of the adaptivity of the human element, who, being goal motivated, stresses himself towards success of 'his' MMS. Ergonomic deficiencies and suboptimality only show up at peak performances—in the form of greater variability[1] in performance, in higher error rates, and in unavoidable increases in the cost of ownership.

'Ergonomically justified design' implies that people taking part in the socio-technical system are being regarded and employed in a personalistic manner, i.e. which takes account of the existence of an individual's character and needs; an approach which, of course, does not imply that each workplace must accommodate all sorts of idiosyncracies. However, a balance must be struck, and the currently accepted view is that at least anthropometric characteristics (sizes, distances, weights, forces) of the workplace must be adaptable to the individual's data and capabilities. This view stems mainly from medical considerations.

The logical extension is that similar requirements hold for psycho-physical and psychological factors inasmuch as they apply to the work in question. This does not mean that human responses to external stimuli cannot be used in a mechanistic sense: it may well be that for reasons of safety, or of quality of the product, and so on, the human element must be trained to respond according to well-defined rules.

System design

The steps involved in 'top-down' design of complex (socio-technical) systems are basically similar to those taken in the design of more modest equipment. Instrument development in the early days of physics provides good examples of design, showing, among other qualities, a high degree of feedback from the user and the manufacturer alike.

Several centuries ago, it was more the rule than the exception that an instrument was conceived and developed at the time it was needed and by the innovator who wanted to make use of it. Until recently, it was common practice to design smaller systems in-house, on a growth basis, adding functions or increasing capacity as required. This technique, known as 'operational development', has advantages over the custom-made approach because of intrinsic feedback in the development stage: the designer is well

---

[1] Occasions of disinterest in the objective also lead to serious mismatch between man and machine, but a fault-tolerant system is not always possible nor the desirable solution.

aware of the capabilities and limitations of procedures and equipment, their interrelations, their effect upon the object to be manipulated, and on the necessary administrative techniques. The sensitivity of every action in relation to the final result can be considered in advance. For the quality improvement of some products the operational development technique is even indispensable.

For complex systems, a design methodology which preferably makes use of computer-aided design methods is desirable. Within a methodical framework it is possible to partition the whole into manageable entities. To do so, one must be able to identify different aspects and structures which describe the relations between the smaller parts. The process of systematic design thus involves the breakdown of the main system function into system processes, maintained by subfunctions, and so on; including thorough analysis of the secondary (with respect to the functions) requirements and of the environment in which the system must operate.

In the analysis phase, during the partitioning into subfunctions and into all sorts of aspects, attention must be paid to the (inter)relations between components and between components and human elements, and to non-functional aspects which are vital to the proper operation of the system, such as maintainability, electromagnetic and chemical susceptibility, and the like. In the synthesis phase, the design concentrates on realization of methods and implementation of means (equipment) such that the assembled whole conforms to the results of the analysis. Care must be taken that a continuous review of design progress is maintained, so as to provide feedback data to enable evaluation of every step and every decision in the design process. In earlier stages of the design, the degree of detail is rather too coarse for correct judgement of such aspects. An undesirable outcome apparent at an intermediate stage must often be traced back to a particular decision at an earlier stage. Consequently, the design alternatives must be carried out to an intermediate stage where rejection of some alternatives is justified and indications are obtained for revision of certain decisions made in preceding stages—see Figure 11.1, which depicts the design process of a measurement system. Iteration is basic to design. Good, or acceptable designs of a product are usually achieved because in those cases the evolution is very gradual: only small changes are introduced and by experienced designers (operational development).

The technical discipline 'systems engineering' and the science of 'systems theory' provide (as yet meagre) methods and rules for an orderly, methodical decomposition of systems into smaller subsystems and partial systems, down to microscopic detail. Their application does not make designing lighter; on the contrary, it implies more work, but the balance is better control over the whole design process.

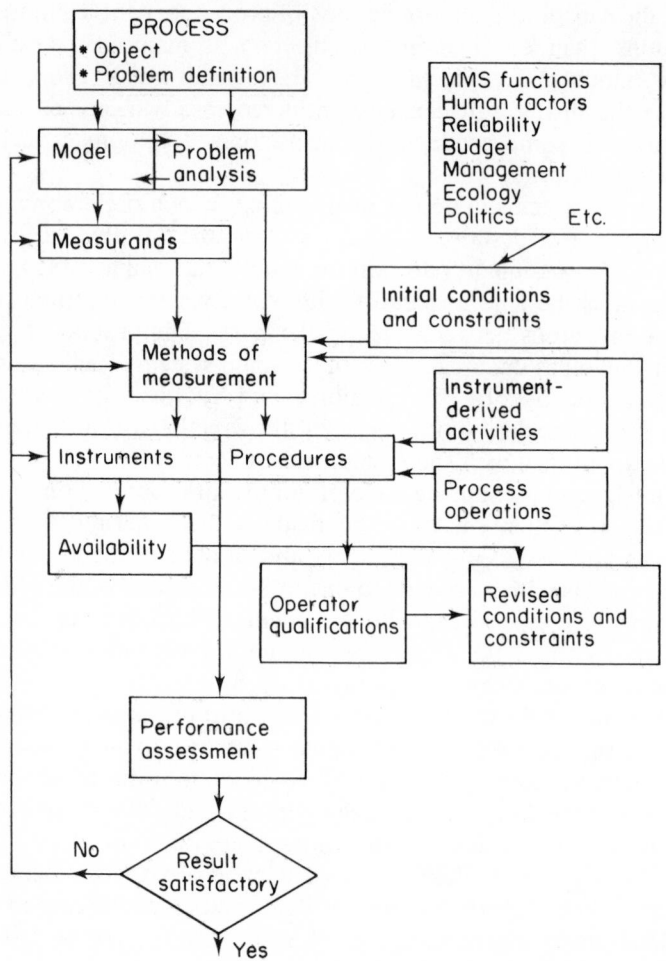

Figure 11.1   Simplified flow diagram of the design of material and procedural means, with
emphasis on aspect assessment (conditions and constraints)

Similar benefits were obtained in the recent past, when the concepts of the
systems approach were mainly focused on the time parameter of the
engineering process and its management (e.g. Hall, 1962; Goode and
Machol, 1968). Planning methods resulted in the decomposition of the
design and production process into phases (for instance: analysis and
exploratory phase; preliminary design and test; production design and

tooling; production; acceptance test and start-up; after-sales service) with the object of controlling budgetary and scheduling aspects.

Recently more insight has been gained into the methods through which engineering design can be systematically pursued (e.g. Asimow, 1968; Hall, 1962; Koller, 1973; Spillers, 1974; Mesarovic *et al.*, 1970; Edwards and Lees, 1973; Kraiss and Moraal, 1976).

Structure of the chapter

The design process can be regarded as a system itself; looking back at his trade, every design engineer will perceive a complex network of decisions which encompasses every aspect, every subsystem, and every activity of the system.

Methodical partitioning must provide room at higher hierarchical levels (coarser detail) for realization of alternative methods or at lower levels (finer detail) for implementation of alternative means. Therefore some generalized system concepts will be reviewed first, after which the methodical design process is briefly described. The chapter concludes with the role of the human factor.

System structures

System reticulation is a highly theoretical subject which, when applied in practical situations, boils down to commonsense techniques 'which one has been doing all along' (Rechtin, 1968). The total process is segmented into smaller elements in order to facilitate the identification of methods and means by which the process can be operated. Study of the partial processes carried out in each separate element shows the relation between methods and means in a diagrammatic structure.

General structures

The partitioning of a system can be done in several ways. For our purpose the most useful and therefore commonly encountered are:

Subsystems: separately identifiable conglomerations of equipment (instruments) with connected (partial) functions which are logically geared to the system's main goal. The hierarchical level of the main system decides what equipment is considered to be a subsystem. At the level of, for example, process instrumentation, a data logger will be considered a subsystem. In relation to the data logger, an A to D converter, a dedicated computer, and a data formatter are subsystems. The A to D converter has only a few functional relations with the other subsystems of the data logger, whereas the data formatter has many, which spread out like a nerve system—yet the data formatter's function is objectively definable.

Partial systems or aspect systems: a generic term denoting networks of each system attribute belonging to *one* aspect. For instance, aspects which every subsystem shares with all others are: supply of power, internal temperature, sensitivity. Also, functions can lead to an aspect network: the A to D converter depends for its operation on a specified sequence of events (internal programme) which must be synchronized with other subsystems; likewise the data formatter and the computer. Another example: the total act of navigating an aircraft can be described as an aspect system, encompassing navigation subsystem *equipment* and the navigation *procedures* for the particular aircraft.

A non-functional attribute which is vital to the system's performance is 'accuracy'. The system's error budget is not a by-product of design. Different contributing system parts are interrelated in this aspect through the model of the error propagation analysis and of the process activity in which the equipment takes part.

Both functional and non-functional aspects' characteristics are frozen into the firmware by design. This is exemplified in Figure 11.2, in which a choice of attributes is grouped together, for example size, weight, power, and cooling, under the heading 'physical aspect' (Shepherd, 1974). The diagram is by no means complete: a software aspect, for example, is most important and should be studied. It still takes experience to know where to stop! But at least this systematic approach helps in formulating (such) questions which, in non-methodical design, do not even arise until too late.

System activities

An aspect of almost overriding functional importance comprises the activities which go on within the system, either autonomous or under control of external stimuli, the amount and type being determined by the particular choice of equipment and the degree of automation. One may distinguish *procedures* and *programmes* or *algorithms*.

Process activities, whether they be manual, mechanical, electronic or otherwise, generally include branching points. If the choice of action or decision at these points is vague, i.e. not fully determined by previous operations in the process (they may also depend on independent internal states, in the system, and external states, in the environment of the system), the composite activity is a *procedure*. The vagueness at the branching points requires some higher-order *decision strategy*, often conveniently provided by a human operator. If the MMS is to be fully controllable, the decision

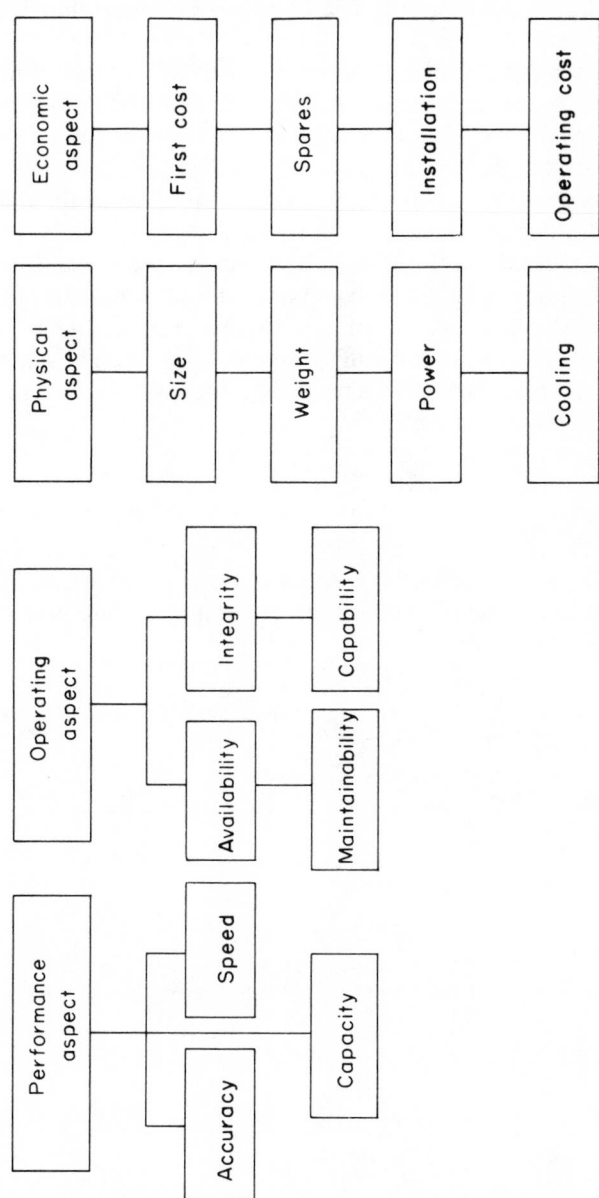

Figure 11.2 Partitioning into aspects

strategies must be considered explicitly in the design of the MMS. We will come back to this later. Sometimes it is possible to implement the strategies in some intelligent machine, capable of taking optimum decisions in fuzzy situations.

If, on the other hand, the choice of action is fully determined by the structure of the system and by the previous operations, the composite activity is a *programme* or *algorithm* instead of a procedure. Procedures, programmes or algorithms are also subsystems in our sense, and quite commonly used in, for example, computer software, but also in many industrial and other activities.

Nowadays, the partitioning into subsystems, partial systems, and system activities is common practice. It has proved to be useful in, *inter alia*, the assignment of tasks to project teams (see Figure 11.3, which shows part of a possible breakdown for an air traffic control design team, each team doing part of the job). Also it helps in the analysis phase, provided of course there is already something to analyse.

Further partitioning

System functions are realized by specific system operations. The sort of operations determine which kinds/type of equipments are going to be used

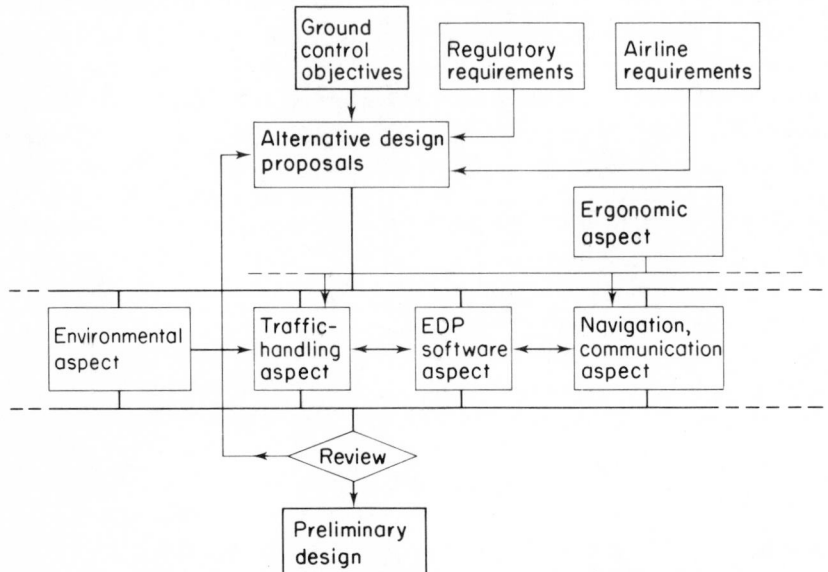

Figure 11.3   Aspect breakdown of ATC design project team

and which activities carry out the chosen processes. To arrive in a methodical fashion at the decisions involved, further decomposition into other kinds of structure (Bosman, 1967) is required, including identification of the interfaces which come into being as a consequence of the partitioning. Many types of structure can be conceived:

— *Functional*, indicating the system functions which follow from the list of requirements;
— *Organic*, describing the system in terms of selected hardware/equipment;
— *Activity*, denoting system activities and their interrelations;
— *Socio-technical*, giving man–machine interrelations;
— *Information*, showing signal/data flow;
— *Social*, describing interhuman relationships;

and so on.

Some are primarily goal-related aspect systems (functional, activity, information) while other structures (organic, socio-technical, social) are, although goal-derived, more concerned with constructional and societal considerations. This chapter is mainly concerned with the first four listed, since these determine to a large extent the MMS capabilities and the understanding of their jobs by the personnel involved: the inner representation (architectonic aspect) whose ergonomic significance is as yet insufficiently understood. In order to meet its objectives, the MMS must accomplish a number of (sub) functions in the system. In this context, *function* is defined as 'mode of action by which it fulfils its purpose' (Fowler and Fowler, 1963), in our case the action being that input variables are transformed into desired output variables. The variables can be matter, energy, and information. Functions are derived directly from the set of requirements, which are expressions of the obligations the system must fulfil. It will be seen that, in the process of system decomposition, the functional structure is much closer to the purpose or system objectives than is the organic structure.

A simple example to explain the notion of function can be derived from a well-known means of transportation: the bicycle. Its function is as said: transportation. To accomplish that function, a number of subfunctions are required. Among others: to carry the weight of the rider irrespective of the bicycle's location or speed, to guide the vehicle along the chosen path, to prevent sideways slippage, to transform torque exerted at the pedals into a linear force, etc. These subfunctions are connected to form a network of subfunctions, the *functional structure*, which accomplishes the main function of the bicycle. It stands to reason that the (human) rider adds the required subfunctions: to supply torque, to maintain equilibrium, to navigate (see Figure 11.4).

In more complicated vehicles one may distinguish a chassis, a number of undercarriages, engine compartments, a body or fuselage, steering system, etc. This grouping of functions according to external conditions or interface criteria (e.g. minimum wiring between boxes, equipment architecture, maintenance aspect, or sheer mechanical necessity) leads to the *organic structure*, derived from the functional structure but better suited for descriptions of the system in terms of selected hardware/equipment and also for anthropometric analysis. Technology and engineering provide the *methods* and *means* to realize each function: the organic structure models the anatomy of that realization.

From this example it will be clear that the blocks in the organic structure are system components or subsystems. These blocks, by analogy, can be termed *organs*, a notion equivalent to that in biosciences, where for example a hand, a liver are organs (physically integrated subsystem), while for example the motoric nerve system is a physically distributed subsystem. The organic structure shows the distribution of equipment according to constructional criteria. The familiar diagram of a computer is depicted with organs instead of functions. It shows the central processing unit, core memory, disc units, a tape reader, etc. For the purpose of explaining internal activities such a diagram is quite useful provided one keeps in mind that the CPU is flexible in a functional sense (it is programmed to perform different functions in the desired sequence), whereas the tape reader's function is single and frozen in the firmware. Such diagrams are organic structures, the function of which can be apprehended only in conjunction with the variable part of the activity structure (in this case the programme). Changing the function means loading a different programme.

Organs are physically integrated sets of components of subsystems, designed for (a small number of) well-defined functions in the total process of the system. In this sense, software modules or subroutines also comply with the definition. These organs are operated through programmes, algorithms or even procedures such that the specified functions are accomplished. It follows that activities (programmes, algorithms, procedures) and the cooperating organs are hierarchically at the same level. The functional structure, being generic for the organic structure, is at the next higher level. The *functional flow diagram* (FFD) is generic for the activity structure of the system. Flow systems characteristically include a causal relevancy between input and output variables (dynamic characteristics of the system). The dependent variables must be controllable in some way and may be (a combination of) information, energy, and matter.

One way to analyse flow systems is by a flow diagram. In Figure 11.5 an example (van Cott and Kinkade, 1972) is given of an FFD (left-hand side, line

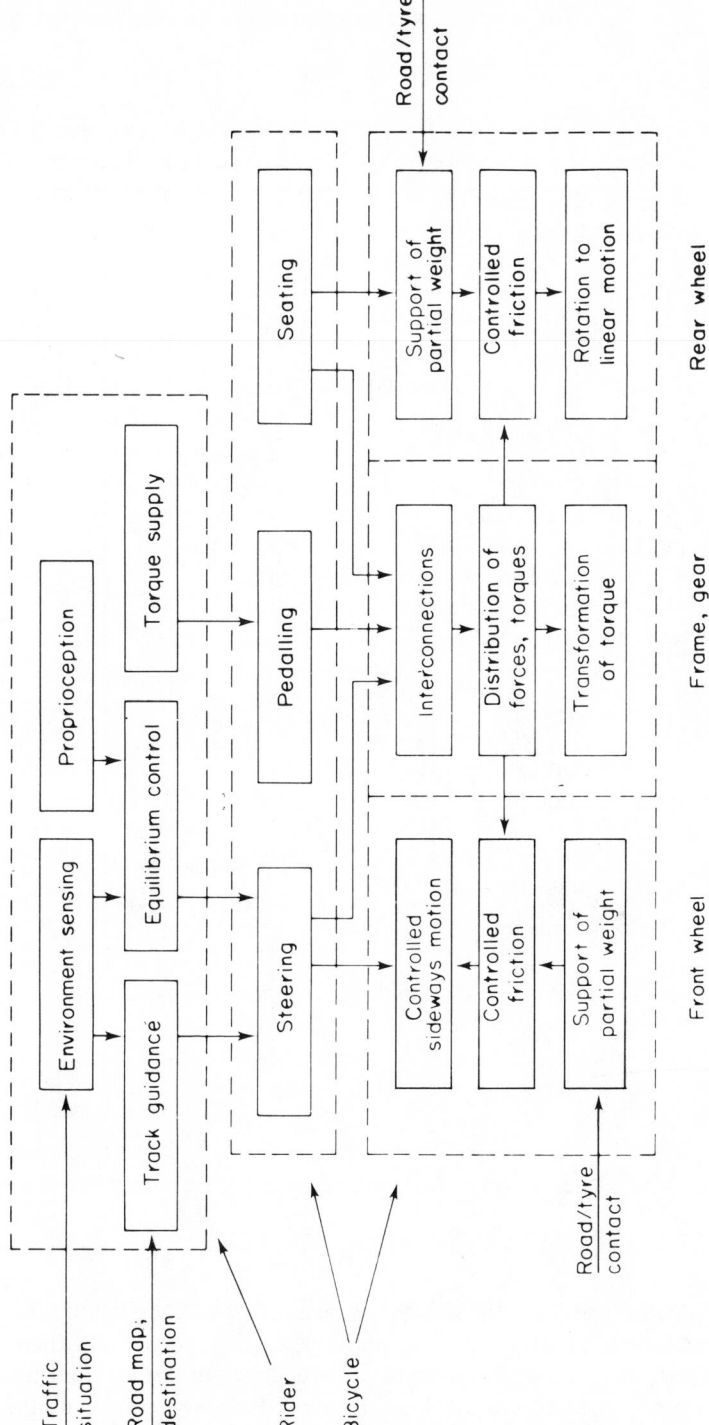

Figure 11.4 Functional structure of bicycle MSS. Boxes show partitions to develop organic structure

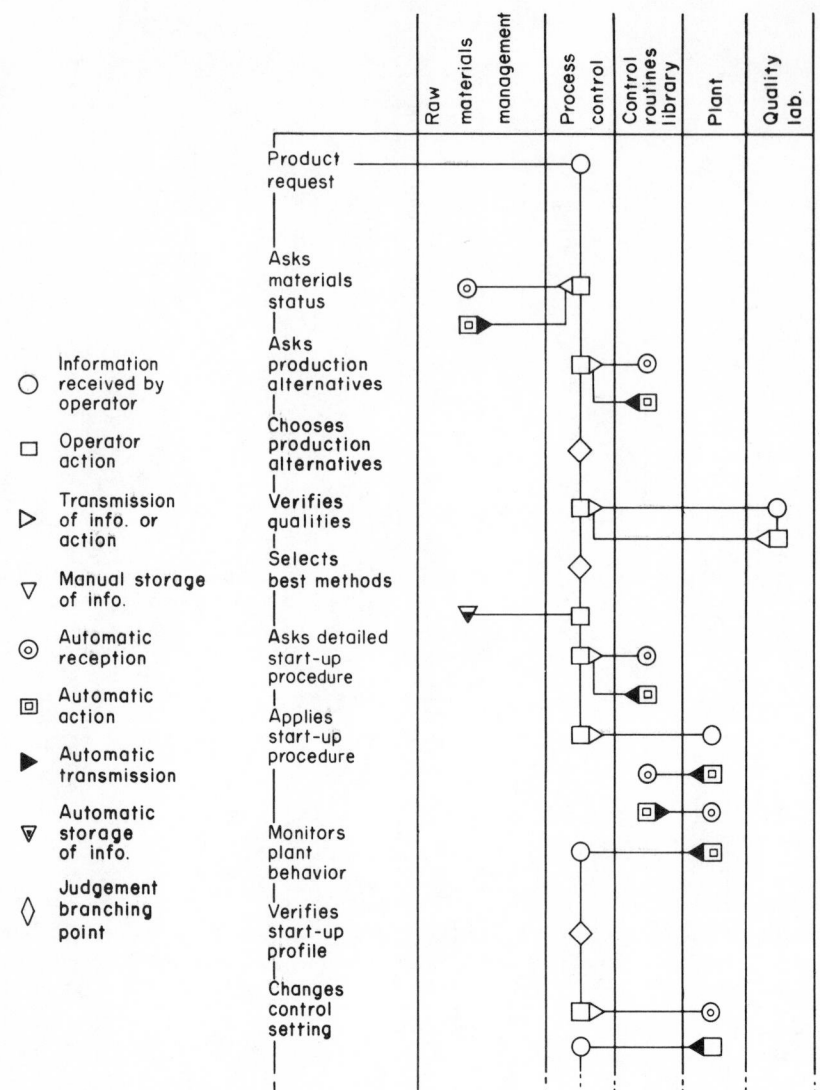

Figure 11.5   Process control procedure flow diagram (activity structure)

structure) of a hypothetical process which may serve to illustrate the method. At the right-hand side is its derivative, the activity structure. Note that there are decision points, marked by diamonds, where the operator applies his own judgement and is free to choose from known alternatives or to end a current procedure in favour of branching into corrective action, depending upon the circumstances.

The other branches originating at these decision points are not shown; this figure describes only one alternative of all possible partial activity structures. If every alternative possesses n m-ary branching points at comparable stages in the process, the total number of alternatives is m^n. For m, n larger than 2 the number quickly outgrows our mental capacity. One should either be in a position to avoid that situation or request the help of a computer which may select, according to a ranking procedure following from a cost function, the four or five most acceptable alternatives. The criteria for choosing one of these alternative (partial) activity structures are sometimes rather soft, necessitating a higher-order decision strategy which takes into account more factors and more actual data than were used in the cost function during the design phase. This is mostly provided for by a human operator, although adaptive activity structures, controlled by a dedicated (learning) computer, are quite feasible. In the latter case, it is clear that the basic concepts for that strategy must be provided by the designer of the system; *mutatis mutandis*, that is also true for the human operator.

The example of Figure 11.5 illustrates that already in the early phases of design FFD analysis can be done in a methodical fashion. Each of the functions shown in this process can be detailed into a composition of (lower-order) functions: every FFD can be used to generate FFDs of lower hierarchical order.

The following example is chosen for its recognizability by a wide spectrum of disciplines rather than for its practical value.

In mechanical construction, a common requirement is the production of holes of specified dimensions. Drilling holes is but one function in a FFD which describes in general terms a construction process. The FFD of the drilling process (also in general terms) is shown in Figure 11.6.

The *function* is: to remove material within specified coordinates. The *methods* applicable to this function are diverse and chosen according to criteria external to the function (aspects), i.e. electron beam cutting, spark cutting, etching, drilling, etc. Suppose the method of drilling is chosen. The *means* to implement the method is to apply rotational energy to a drill, which is positioned within a specified set of coordinates, the energy and position being controlled by information derived from the progress of the drilling process.

It can be seen that material means (positioning), energy, and information cooperate to produce the hole. Figure 11.6 depicts the functional flow diagram of this process, under the assumption that a human controls the process by dosage of drilling power (perception of chip characteristics) and by workpiece positioning. As such, it is a feedback process, carried out by a single individual.

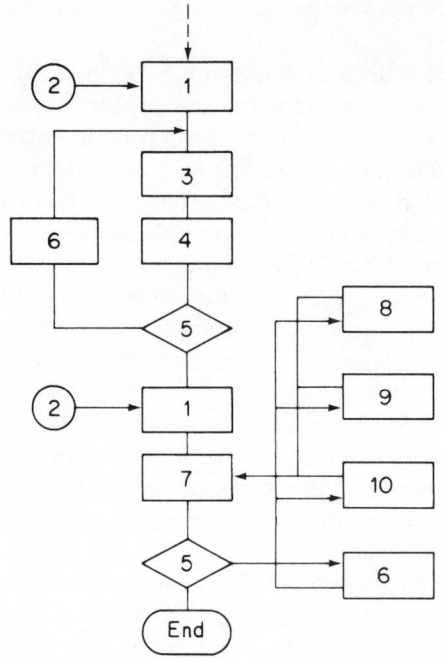

1   DETERMINATION OF ACTION
    PARAMETERS
2   QUALITY CRITERIA
3   FITTING OF WORKPIECE
4   POSITIONING OF DRILLS
5   EVALUATION OF RESULT
6   DETERMINATION OF ACTION
    PARAMETERS
7   INITIATION OF DRILLING
8   APPLICATION OF ROTATION
    POWER
9   TRANSLATION OF DRILLS
10 LUBRICATION

Figure 11.6   FFD of drilling a hole

Positioning is done in the initial phase 1 through 4, the result being reviewed in 6 after the branching point 5. In the next phase rotational energy is applied by translating the drill in the material, with the review process having a more continuous character than in the case of positioning. The operator can control applied power through control of lubrication and of rotation and translation speed.

The automated version would most probably employ feedforward control instead for power and position.

After activity analysis it will become apparent that the activities can be grouped into four categories:

(a)   To be performed by equipment only, in some cases monitored;
(b)   To be carried out by human operators only;
(c)   To be done by either equipment or human operators, the choice being dependent on external conditions and/or constraints, possibly even being programmable;
(d)   Can only be done through cooperation between equipment and human operators.

The activity structure is thus the network of all activities taking place in material means (through programmes and algorithms) and induced by personnel (through procedures), structured in combinatorial and sequential fashion to meet particular needs and specified functions.

The *socio-technical* structure is the network of interrelations between human activities and equipment activities. It can be derived only when the activity allocation and the choice of the organic structure have been made. To this end one should first supplement the process-oriented procedures with equipment support activities, and possibly with additional tasks required by management. The resulting tasks must also be distributed in time so that the derived jobs comply with biological and social needs and constraints.

Perceptive aspects of system structures

With the aid of the FFD the designer may, as early as the exploratory phase, obtain insight into the required activities of both a procedural and material nature.

On the other hand, in the operational system the activity structure reveals to the technical manager and to the operator much of the subsystem's FFDs which together, in relation to one another and to the environment, constitute the *functional architecture* of the instrumentation. This manifests itself by, for example, the length of procedural activities between decision points, by the number of decisions per unit of time, by possible combinations and sequences of operations, by the required personnel selection, instruction, and training, by work–rest cycles, etc. The functional architecture is important for the operator's understanding of the system's operation in the process, for his inner representation or 'mental model' (Mackay, 1963; Bainbridge, 1969) of it.

The other constituent of the operator's perception of the system is the *sensory architecture* derived from the organic structure. He sees controls and displays in a particular setting, some of which are vital windows to functions important in the process control operations. Different shapes, sizes, colours, and other presentation aspects assist in the identification of partial functions of controls and displays, and facilitate their connection to procedures. Figure 11.7 is illustrative in this respect. It is a design aspect not to be taken lightly to match the designer's concept of the operation of the system to a profile of the most likely inner representation of the average operator. Thus, a new design for an equipment or system should not only start with questions about the process, but must include some about type of operators and what is known about their behaviour and previous experience as system components (Rijnsdorp, 1976; Rijnsdorp and Rouse, 1977).

To the maintenance crew, however, the organic structure means no more than a conglomeration of boxes, cabling, calibration controls, status checks,

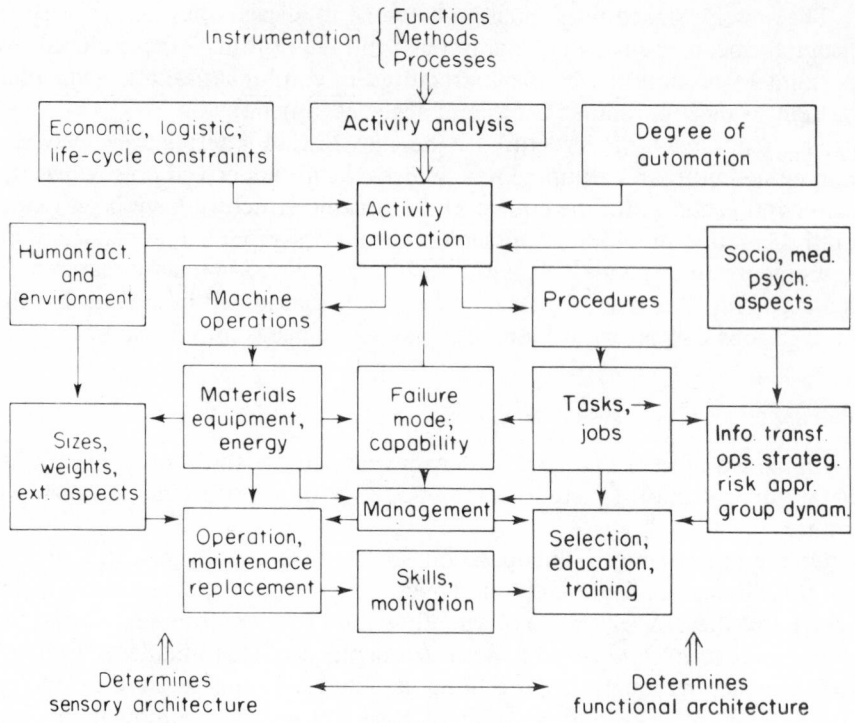

Figure 11.7　General synthesis diagram with activity allocation in the central position

etc., because normally they are not familiar with the details of the activity structure pertaining to that equipment or system.

The distribution of activities over system parts and operators greatly determines the functional architecture. That distribution results from a design process called activity allocation, which happens very early in the design effort, at the coarse level of detail, when it is hardly possible to foresee all consequences of each decision. That is the reason why the inclusion of ergonomic considerations is already of vital importance in the early phases of design; and also why the design process must be of an iterative nature. In order to arrive at the decisions involved in the activity allocation in a methodical fashion, we will now turn to the design process itself.

Structuring the design process

Introduction

It is essential that at the start of a design effort the MMS objectives are precisely known to the future user; to the extent that these are vague, the resulting system will be imprecise. Also, that the user has sufficient wisdom to refrain from defining for himself the hardware configuration to be produced by the designer. Further, it is a 'must' that the user and the designer together translate the objectives into MMS requirements, and that they agree on initial conditions and constraints of a general nature to be applied to the design; for example cost ceiling, environmental factors, human factors, ethical and aesthetic considerations, etc. (see Figure 11.1).

A number of decomposition techniques are applicable. The time span of development and design is divided into several phases (such as exploratory phase, analysis and planning, preliminary design, detailed design, test phase, and operational introduction: Goode and Machol, 1957; Hill, 1970; Asimow, 1968. For the realization of a complex function a team of designers is required, working to project management methods. One way of partitioning the problem is into different aspects as exemplified in Figure 11.3. The team is split up into groups each handling, for example, the aspects of: software, electronics, packaging and integration, user procedures, human factors, etc. With such a number of people with diverse responsibilities working on the same functions, one must be able to structure the design process so that each is aware of the stage of development of his own team in relation to the progress of others and can assess the impact of design decisions in his area of responsibility on the freedom of choice of his fellow designers. Moreover, critical path constraints in the management network may enforce coordination of progress in the separate aspects.

A logical way of progress phasing is the sequence:

Objectives (system's goal)
Requirements (obtained after problem analysis)
Functions (derived from requirements)
Methods (to realize each function)
Means (implementation of methods)

For example, in a computing system the result of the activities of peripherals, CPU, and storage is processed data. A reason to make use of that system could be to play an intellectual game, or to calculate pay cheques, or to solve a mathematical problem, etc. These trivial examples illustrate how different objectives can be achieved by the same *function* (see upper part of Figure 11.8).

Every function is carried out by a suitable combination of human and material activities, i.e. each function is realized by a chosen *method*.

Generally, there is not a one-to-one relationship between function and method, because often several methods are available for the realization of a

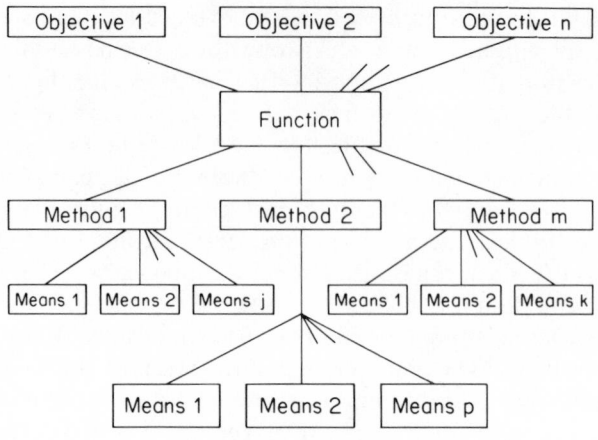

Figure 11.8   Realization tree with 'function' in the central position

function. For instance, several methods are available for the transducing of pressure, for example the use of:

— The elasticity of a material (membrane)
— The change of stiffness of a resonator (string or crystal)
— Gravity (liquid manometer, dead weight tester)
— Gas-law
— Acoustics

Also, transducers may be obtained for different types of output, accuracy classes, and environments. The relations between functions and aspects are methodically explored. The aspects serve in this respect as conditions and constraints for the choice of a method of measurement and/or an instrument as shown in Figure 11.1.

In the same way as the analysis of the transduction function, other functions of a projected instrumentation must be studied. For instance, the function 'output format control' can be achieved by the method of a scheduled sequence, producing data irrespective of their values, or by an on-demand interrupt method, producing data including identlabel only when they are relevant to their end-user, or by other functionally equivalent methods.

Once a particular method is selected, there is a variety of equipment *and* procedures (*means*) for its implementation. Thus the methodical analysis can lead to a tree with a very wide base, the final selection of the configuration being derived from those combinations of aspects (conditions and constraints, Figure 11.1) which best satisfy the external criteria (cost, maintainability, reliability, etc.). In the example above, the subfunction 'addressing' in the method of scheduled sequence can be realized by, for example, a digital method assigning address bits to each variable; this in turn may be implemented by a binary counter, or by a shift register combined with a look-up table, or other methodically equivalent means.

On the basis of the analysis of functions and methods, which leads in the exploratory phase to different instrumentation configurations, the instrumentation engineer starts to study the resulting alternative FFDs, leading to lists of system activities (equipment and software and operators).

In order to keep the design effort within acceptable proportions, one must be very selective in the upper levels of analysis (requirement and function, possibly also method), but with preknowledge of the possible effects at lower levels—especially in the area of human activities (e.g. degree of automation). In terms of system structures, comparable levels are as given in the diagram below.

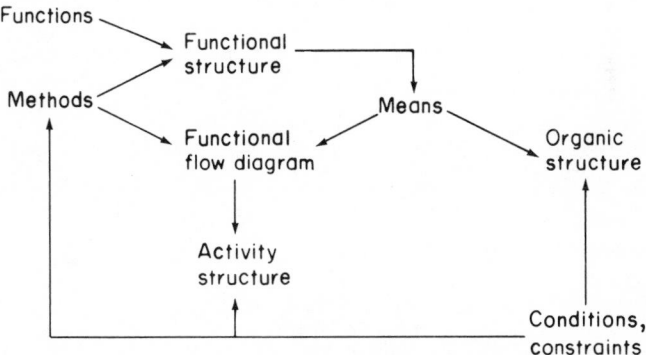

In Figure 11.8 a realization tree is depicted with the notion of 'function' in the central position. This situation occurs for mass-produced 'general purpose' equipment. In that case the designer is usually unaware of the precise objectives of its user. Most computers, for example, belong to this class. Conversely, in the design of goal-directed or dedicated equipment, this realization tree should only have the precisely known system objectives as its central sources; it becomes the equipment design tree. The tree characteristic is usually removed from block diagrams of design processes, because they would quickly become unwieldy. A typical example, but extremely simplified in order to emphasize the basic 'objectives–requirements–functions–methods–means' sequence, is shown in Figure 11.9. Reiteration, so characteristic of every creative process, is symbolized by the many feedback loops.

Reiteration occurs by comparison with the alternatives of other partial functions, by comparing against initial design conditions, by invoking

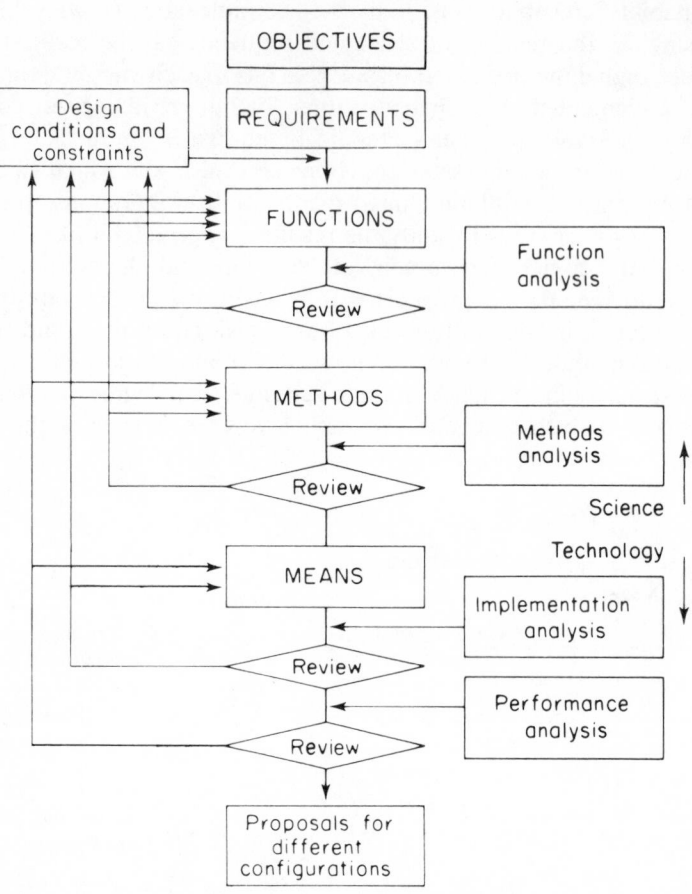

Figure 11.9   Reiterative design loops

practical reasoning. A well-known example of the impact of the choice of means is the different computer architecture obtained with either core storage or disc storage. And who does not dream of the 'hours to go before breakdown' indicator? It would make the design of maintenance programmes so much simpler. Theoretically, the choice of functions (leading to the functional structure, see earlier) is the more sensitive with respect to reiteration. In the practical situation, where design experience is already available, not the functional level but the choice of methods and means is often the more vulnerable to the quality of decisions.

The choice of methods and means automatically includes the allocation of process activities to operators and to equipment (see Figure 11.7).

Instrumentation systems fail to produce good results if the activities involved in their application are incorrectly carried out.

In many cases, the quality can be made to match the required integrity[1] of equipment operations, although this may not comply with cost considerations. Human procedural integrity is less controllable. When a low error rate is essential, adequate activity analysis (see, e.g., van Cott and Kinkade, 1972) *must* be part of the system design, and may prove to be an important source for iteration.

The human factor

One major output of activity analysis, in addition to the performance specification, is (are) the job description(s) which state(s) the required skills, the education, and the general background of the suitable type of operator. There is a trend in instrumentation to become more and more complex, which traditionally requires higher education of the personnel involved. It is the opinion of the author that we must break with this tradition and that designers should aim at producing man–machine systems such that operators with modest general backgrounds and intelligence can fully understand their operation and maintenance.

Ideally, the operator must be able to interpret *his* experience of the MMS behaviour and find satisfaction in his day's work through adequate use of his mental and physical abilities. Today we are far distant from that ideal. It is barely possible to measure the mental and physical load of a complicated task, let alone such vague notions as satisfaction, irritation, boredom, etc. To make such notions technically operational, integrated research of various disciplines such as physics and the behavioural and management sciences is necessary. However, let the gaps in our knowledge be no reason for neglecting these areas. More often than not, one may succeed in designing better instrumentation if available data about the human component are applied as early as the problem definition phase of the project. And that may be especially true when man–machine systems are designed for different ethnic groups, both in terms of functional and of sensory architecture.

In any case, the allocation of activities to man, the grouping of activities into procedures, then into tasks and into jobs, inevitably invokes the discipline of organization in the project, as shown in Figure 11.7. To the operator, every action in his task is not only a source of error, but also a part of a meaningful sequence which relates to the process behaviour. This enables him to evaluate the acquired data (quality monitor function) in order to detect possible malfunctioning. Another important source of human error

[1] Integrity of the operations of the instrumentation system is proportional to the probability that its performance remains within tolerance limits notwithstanding small changes in equipment and/or procedures.

stems from misunderstanding. An operator builds up his (own) mental model through assimilation of the factors which make up the functional and sensory architecture of the instrumentation. To the extent that this information is insufficient or not comprehended, he will supplement it with his own notions. Consequently, his mental model is strictly individual and may deviate considerably from the designer's image of the modal operator. Good, transparent functional and socio-technical structures are a great help in coping with unforeseen situations; one cannot rely solely on failure mode analysis and thorough training of the operator. Because there is as yet no way to measure the properties of mental models, the best way to achieve some compatibility between the instrumentation's architecture and the operator's internal representation of it is, already in the exploratory phase but also in later stages, to ask operators of similar systems to give their opinion of projected jobs.

The choice of methods and means implies the allocation of process activities to equipment and to human operators. Some of these can be easily mechanized; others, like routine control tasks, are amenable to automation. When these are not too dull, or do not require too much of the capacity of the human being, they can be assigned to an operator. If, on the contrary, the job is (becomes) very demanding, as in critical situations, training in (longer) overlearned patterns increases the operator's reaction speed and his information-handling capacity by a factor of 2. It has also been shown that mental overload leads to partial slippage in (lower priority) secondary tasks (Kalsbeek, 1967). In general, one sees a shift in MMS with a higher degree of automation from motoric towards mental skills; it is questionable whether this trend must be accepted or should rather be challenged according to criteria of job satisfaction and of skill reward, but also of cost effectiveness.

The performance of the MMS depends on the weighted sums of the capabilities of both the technological part of the system and the human operator, involving the degree of matching between system properties (equipment, process activities) and human factors such as inner representation or mental model, information-handling capacity, motoric and mental skills, selection and (over) training, briefly covered in the following section. Peculiarities such as stereotypes, illusions, subjective interpretations, psychophysical abilities and limitations, physiological constraints, and anthropometric requirements are *inter alia* treated in van Cott and Kinkade (1972), Krais and Moraal (1976) and Singleton *et al.* (1971).

In the activity analysis and the subsequent activity allocation (Figure 11.7), the aspects of timing, speed, accuracy, computation, energy expenditure, strength, etc., become tangible in the sense that not only do sequential and combinatorial relations appear in writing but, most important, can be enumerated quantitatively. Those activities with aspects which lie within the human aspects profile can be allocated to man.

Traditionally, the operator's ability to perform in a specified task is measured following an outline of aspects such as listed in Table 11.1. This table is suggested by V. David Hopkin (1970), and gives an excellent exposé of human factors in the ground control of aircraft.

Table 11.1

1. Biographical requirements: age, sex, nationality, experience in job, other relevant experience
2. Physical and physiological requirements: attainment of medical standards, general health, physique, strength, endurance of or resistance to fatigue, tolerance of minor physical environmental stresses, adaptability to work/rest cycles
3. Sensory requirements: auditory, visual, tactile, kinaesthetic, sensory interactions
4. Perceptual requirements: speed, accuracy, and ability to discriminate in each sense modality
5. Information-processing requirements
6. Psychomotor requirements: muscular coordination, fine coordination, dexterity, manipulative abilities, compatibility between stimulus and response
7. Verbal requirements: language(s) spoken and understood, verbal fluency, clarity of expression
8. Knowledge and skill requirements: academic knowledge, practical knowledge, training records, ability to apply knowledge, skill in job procedures, practical judgement
9. Educational requirements: basic educational qualifications, additional qualifications, recent attainments, courses attended, future education
10. Mental and cognitive requirements: general intelligence, verbal, numerical, spatial and mechanical abilities and aptitudes, reasoning, short-term and long-term memory, ability to innovate, flexibility of thinking, ability to learn from experience, ability to forget or discard inappropriate behaviour
11. Clerical requirements: speed and accuracy in associated sensory, perceptual, intellectual and psychomotor functions, preparation and use of job aids
12. Personal requirements: general personality, specific personality traits, personality profile, personal appearance and habits
13. Social requirements: ability as a team member, tact, leadership, morale, attitudes towards superiors, attitudes towards subordinates
14. Interests and motivational requirements: activities and behaviour which reflect job interest and job satisfaction, need for challenge and effort in job, opportunities to use skills
15. Emotional requirements: emotional stability, perseverance, tolerance of varying workload conditions, response to stress, response to boredom

Since these aspects are, within the boundaries of their variability, fixed data, it is necessary to compare the required level of effort against the human endurance profile. Some human factors handbooks (e.g. van Cott and Kinkade, 1972) provide such data. Endurance is highly variable but also adaptive: mathematical models as, for example, transfer functions most account for such non-linearity in actual stress conditions. Also, man could be conceived of as a cybernetic being, with the ability to make decisions under ambiguous circumstances, to learn and to change his mind, to test and to validate or to falsify, to judge usefulness, each change of state adding to his experience. Psychic stimulus/response (SR) models shall be tested particularly in simulated environments with various conditions of (goal-derived) mental and physical load. The operator's activity can be far more complicated than the designer's activity analysis predicted, and every action is an opportunity for error. Human reliability has thus become the object of study (Meister, 1971; Swain, 1970).

In a job such as process controller, man will use his own sensors and his observation of indicators to control both the inner and the outer loop. His coordinative and social attributes are called upon when other people, operative in the process, are also involved in the task. A generalized diagram of factors contributing to man in a control task is depicted in Figure 11.10. Three basic areas can be recognized—the general aspects of workplace, job and education—and the loops 1 and 2 (see also Figure 11.12).

The second loop has received the most attention for the last thirty years (see e.g., McRuer and Krendel, 1974; Baron and Kleinman, 1968) and is known as the control loop or 'inner loop'. Governing parameters are plant dynamics and physiological speed limitations (neuro-muscular dynamics and so on). The first loop (decision loop or 'outer loop') is of a much more complicated nature, involving cognitive abilities and limitations of the human factor (see e.g., Sheridan and Johannsen, 1976; Rouse, 1972) which are process-dependent.

Research results in this area, of limited availability to designers, and the few data there are fall short of the requirements concerning optimization of MMS capability through better use of human resources. It would be highly advisable to direct more research to the cognitive elements contributing to the success of the process, but in an engineering, quantitative fashion.

The third area is that of the educated, trained, still learning human processor with directives as shown in the double lined box; using short-term and long-term memories, psychomotor skills etc., and all under the influence of human reactions, needs, and constraints such as 'fright factor, motivation, overload, adaptability, habits'. Education and his individual view of his own role, largely determine the operator's 'internal representation' of his work (Bainbridge, 1969). Being highly personal, it can deviate considerably from the designer's image of the work, the control room, and the operator. Consequently, designers must keep in close touch with operators so as not to stretch their adaptability too much, and similarly 'non-natural' parts of the operator's task must be (over) trained to the extent that the internal representations of all operators taking part in the process become congruent as required in high-stress situations. Good, transparent functional and socio-technical structures of the MMS are a great help to cope with unforeseen situations: one cannot depend solely on thorough training and briefing of the

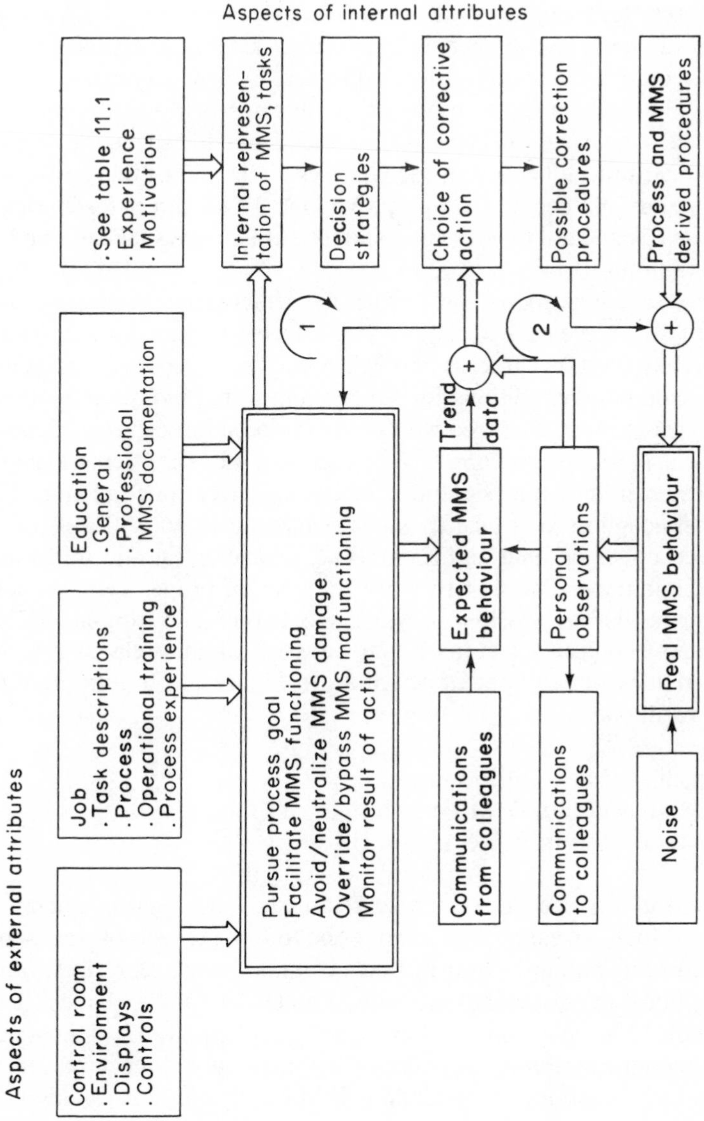

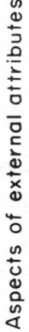

Figure 11.10 The operator as the controller in two simultaneous loops : one cognitive, one dynamic

operator. Moreover, in a well-designed MMS the learning period will be shorter.

Decision strategies are closely connected to the operator's internal representation of the MMS. Generally, they are formed during instruction and exercises, and improved through process experience. In turn, they determine the choice of a specific (corrective) action, given feedback and trend data on the behaviour of the process. As the control experience increases, the operator relies more on predictions of trended data for his choice of corrective action. There is also a tendency to relieve the workload in this manner during periods of mental stress. This can be dangerous when perceptory illusions occur. In highly dynamic MMS direct feedback from instruments is mandatory; both tasks and instruments must be designed with this factor in mind.

Corrective action is applied only when the actual situation is considered to be deviating from the desired datum. It is effected in an open-loop fashion, exemplified by the box 'Process and MMS derived procedures', which also would include basic conditions for functioning of the biological system.

Both open-loop and corrective (decision loop) procedures consist of sequences of partial programmes, their type and sequence being adapted to what is needed in the actual situation, for example according to Figure 11.11.

Maximal adaption to critical dynamic behaviour (Figure 11.12) can only be achieved under optimal anthropometric, physiological and information-handling conditions. The first two are a matter of proper controls design, workload distribution, work–rest cycles, etc. (Kraiss and Moraal, 1976).

The socio-technical structure is the network of interrelations between human activities and equipment activities. In Figure 11.10, *inter alia*, three tasks are mentioned:

— Facilitate MMS functioning
— Avoid/neutralize MMS damage
— Override/bypass MMS malfunctioning

which depend on the socio-technical structure as regards quality and efficiency. Much research has been done on the effect of the sensory architecture on human performance. Optimal distances, sizes, angles, weights; optical characteristics etc. are compiled in handbooks (e.g. Goode and Machol, 1976; van Cott and Kinkade 1972; Fogel, 1967). The literature is less abundant concerning aspects of the functional architecture, although this is most important to maintain MMS *capability* at a high level notwithstanding equipment damage and/or malfunctioning. When equipment faults are 'manipulable', the degradation of capability is controllable by the operator. Hence, it pays to structure both equipment and activities to provide for this capacity as much as possible.

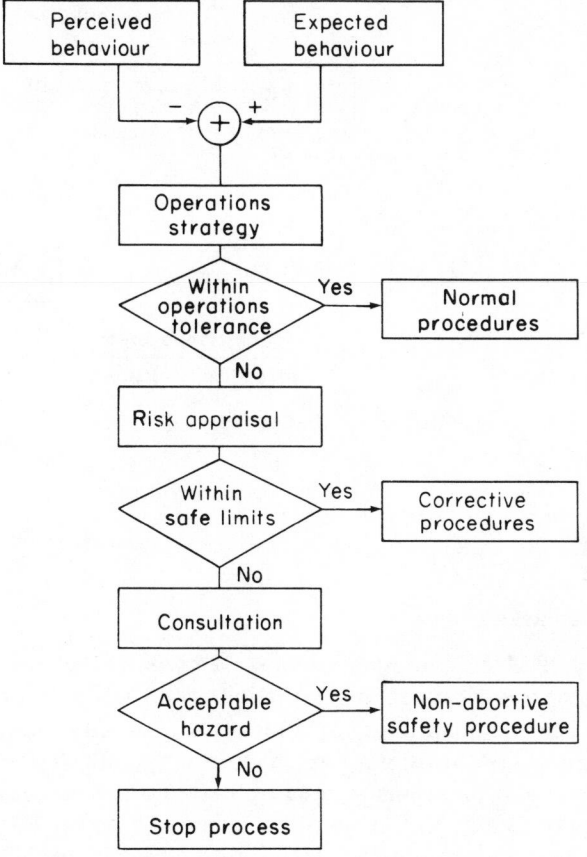

Figure 11.11   Cognitive control procedure alternatives

Capability can be defined as the probability that the MMS will, in a given state and at the proper time, carry out the required design function. This notion is not to be confused with availability, which presents a picture of the condition in which the system is likely to be found at the beginning of process operation. Capability is a measure of the ability of the MSS to achieve with reasonable effort the process objectives and specifically accounts for the performance spectrum of a system (WSEIAC, 1965). A full capability assessment (see Figure 11.7) requires failure mode analysis for every function, taking into account both the failure mechanisms and associated failure rates of the equipment and of the projected human error models (e.g. Swain, 1970; Meister, 1971) applicable to the designed tasks for each of the built-in alternative ways to achieve that function.

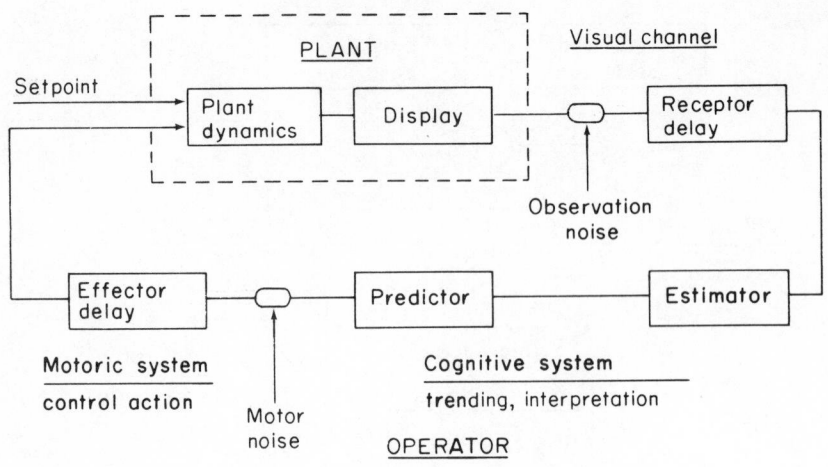

Figure 11.12   Dynamic loop general description

Concluding remarks

In the design of MMS a most important factor is the balance obtained in matching the operator's inner representation of the MMS, its objectives and his power to achieve them to the MMS functional architecture as conceived by the designer. The functional architecture is largely determined in the definition phase and preliminary design phase of a project, and with it the operator's opportunities to function as a cybernetic being. For that reason ergonomic considerations must guide the conception of new MMS and parts thereof. The sensory architecture is mainly determined in the realization phase; anthropometric, physiological and information-handling requirements must be met.

Behavioural sciences were not conceived as a design tool, so one must not wonder why they do not supply sufficient facts and quantified data. Currently, modern technology can provide means for obtaining valid behavioural data in (adequate) simulators.

However, such data still tend to be difficult for the technical designer or the technical project team to assimilate. Also, feedback from actual operators' experience provides important qualitative information. Such procedures, however, are ill-suited to computer-assisted design methods. Thus it will be in the interest of the design community to direct part of its research potential to the non-technical man–machine interface aspects with the aim of obtaining harder, useable data leading to less fuzzy elements in the MMS capability matrix.

Note

The work reported here was started more than a decade ago at the Electronics Laboratory of the National Aerospace Research Institute, Amsterdam; the author wishes to thank his colleagues there for their contribution in the evaluation of applications of the methodology. It was carried on in the chair of Measurement Science and Instrumentation at the author's present address; he is much indebted to prof. ir. J. E. Rijnsdorp, dr. J. W. H. Kalsbeek and ir. F. W. Umbach of the THT Ergonomics Group and to Professor L. Finkelstein of City University, London, with whom he had many fruitful discussions on the subject.

The chapter is a development of the article by the author published in *Journal of Physics E: Scientific Instruments,* **11,** 97–105 (1978).

References

Asimow, M. (1968). *Introduction to Design.* Prentice Hall, Englewood Cliffs, New Jersey.

Bainbridge, L. (1969). The nature of the mental model in process control. Int. Symposium on Man Machine Systems, Cambridge, England.

Baron, S., and Kleinman, D. L. (1968). *Application of Optimal Control Theory to the Prediction of Human Performance in a Complex Task.* Bolt, Beranek, Newman, Cambridge, Mass, USA.

Bosman, D. (1967). Structural aspects of measurement and control systems (in Dutch). NLR Interne Notitie 67,01, Amsterdam.

van Cott, H. P., and Kinkade, R. G. (1972). Human engineering guide to equipment design. Army Navy Air Force, Washington, DC.

Edwards, E., and Lees, F. P. (1973). *The Human Operator in Process Control.* Taylor and Francis, London.

Fogel, L. J. (1967). *Human Information Processing.* Prentice Hall, Englewood Cliffs, New Jersey.

Fowler, H. W., and Fowler, F. G. (1963). *The Concise Oxford Dictionary of Current English.* Oxford University Press, London.

Goode, H. H., and Machol, R. E. (1957). *Systems Engineering.* McGraw Hill, New York.

Hall, A. D. (1962). *A Methodology for Systems Engineering.* van Nostrand, Princeton, New Jersey.

Hill, L. S. (1970). Systems engineering in perspective. *IEEE Transactions on Engineering Management,* **EM-17,** No. 4, November.

Hopkin, V. D. (1970). Human factors in the ground control of aircraft. AGARDograph 142.

Kalsbeek, J. W. H. (1970). Objective measurement of mental work load; possible applications to the flight task. AGARD Conf. Proc. CP 55. Problems of the cockpit environment.

Koller, R. (1973). Eine algorithmisch-physikalisch orientierte Konstruktions-methodik. *VDI Zeitschrift* **115,** Nos. 2, 4, 10, 13.

Kraiss, K.-F., and Moraal, J. (1976). *Introduction to Human Engineering,* Verlag TUV Rheinland, Cologne.

Mackay, D. M. (1963). Internal representation of the external world. AGARD-AVP Symposium on Natural and Artificial Logic Processors, Athens, Greece.

McRuer, D. T., and Krendel, E. S. (1974). Mathematical models of human pilot behaviour. AGARDograph 188.

Meister, D. (1971). Comparative analysis of human reliability models. Bunker Ramo Corp., NTIS: AD 734432.

Mesarovic, M. D. *et al.* (1970). *Theory of Hierarchical Multilevel Systems.* Academic Press, New York.

Rechtin, E. (1968). Systems engineering—but isn't that what I've been doing all along? *Astronautics and Aeronautics,* **6,** No. 6, June.

Rijnsdorp, J. E. (1976). Man–machine communication in computerized chemical plants. European Federation of Chemical Engineering, Firenze, Italy.

Rijnsdorp, J. E., and Rouse, W. B. (1977). Design of man–machine interfaces in process control. *IFAC Symposium on Digital Computer Applications to Process Control.* North Holland, Amsterdam.

Rouse, W. B. (1972). Cognitive sources of sub-optimal human prediction. Ph.D. Thesis, MIT, Massachusetts.

Shepherd, J. T. (1974). The influence of avionic system requirement on airborne computer design. In *Real Time Computer Based Systems,* AGARD Conf. Proc. CP-149.

Sheridan, Th. B., and Johannsen, (Eds) (1976). Monitoring behaviour and supervisory control. Internal symposium, March 6–12, Beichtesgaden, Germany.

Singleton, W. T., *et al.* (1971). *Measurement of Man at Work.* Taylor, London.

Spillers, W. R. (Ed.) (1974). *Symposium on Basic Questions of Design Theory.* North-Holland, Amsterdam, Oxford.

Swain, A. D. (1970). Development of human error data bank. Sandia Lab., Report SC-R-70-4286.

WSEIAC, Weapon System Effectiveness Industry Advisory Committee (1965). Vol. 2 of final report of task group II, AFSC-TR-65-2.

Stress, Work Design, and Productivity
Edited by E. N. Corlett and J. Richardson
© 1981 John Wiley & Sons Ltd

Chapter 12

Case Studies in Job Design for Information Processing Tasks

K. D. Eason
University of Technology, Loughborough, UK
and
R. G. Sell
Department of Employment, London, UK

Introduction

Information technology is increasingly the major force which shapes the jobs of workers of all kinds and at all levels. With the introduction of microprocessors more people are becoming concerned with the processing of information as it becomes cheaper to provide it more widely. This change in technology is also giving more choice to those concerned with the design of jobs, in that they can either create jobs with an emphasis on division of labour or build up integrated jobs which have a variety of different tasks.

All too often, information processing tasks are stressful because, although the computer may free people from much routine repetitive paper-handling, it can create jobs which fall below the standards of good job design. This may especially occur at intermediate stages in the development of information processing systems. Frequent problems are jobs which lack variety, autonomy, and meaning, which involve passive monitoring of computer activity or pacing by the computer. These problems are not, however, inevitable; the computer can also be used to enrich jobs. Computer technology is not deterministic, and even systems designed for similar purposes often have very different effects upon jobs.

The purpose of this chapter is to illustrate by means of case studies some of the options available and how the flexibility of computer technology can be used to actively seek good job design. We begin by examining how two systems with a similar purpose adopted very different philosophies, with very different job design consequences. Thereafter we discuss how one organization tackled a major on-line computerization project and how the

195

flexibility of the technology can be harnessed to make explicit job design decisions.

Job design consequences of data input in on-line systems

In the traditional commercial data processing system one of the key problems has been the volume of data input that was required. This has had profound job design consequences for the small army of data preparation clerks who have to punch data in batch processing systems. However, it is perhaps wiser to look to the future and examine the data input issues associated with on-line computer systems. Let us, therefore, examine how two such systems handled data input.

A centralized data input system

This system was operated by a chain of retail stores with outlets throughout the country. The system handled all delivery notes received by the chain of stores for goods received. All delivery notes were sent to a central office located near the computer centre, where they were checked by one team of clerks and then passed to another team, who input the data to the computer (see Figure 12.1). Each clerk had a visual display terminal and worked through a short, repetitive routine to input each delivery note. This had to be a manual input process because delivery notes were printed in different forms or were handwritten, and encoding them into the standard format for entry to the system required human pattern recognition capabilities.

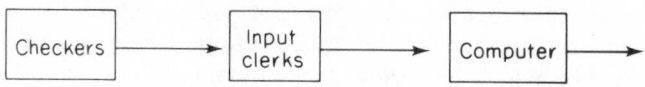

Figure 12.1   A centralized data input system

Upon investigation, it was found that the terminal operators were, at best, apathetic to their work and, at worst, felt controlled and paced by the computer. The system could detect errors in input and request their correction and it could also monitor input rates and produce regular performance measures for the management in terms of quantity and quality. The system was regarded as a considerable financial success by management in the short-term way that it is usually measured, but the company found it difficult to retain staff and had to employ part-time labour with the consequent but less tangible costs that that invoives.

A decentralized data input system

This system was operated by the distribution side of a large food manufacturing organization with a chain of depots for distributing goods to local shops. Each depot was supplied with one or more terminals and it was the responsibility of the depot to input orders received daily which the computer converted into invoices, picking lists, etc. (see Figure 12.2). Each depot organized its own allocation of duties and although data input was largely the responsibility of one person, that person was also involved in many other clerical functions associated with the processing of orders from customers.

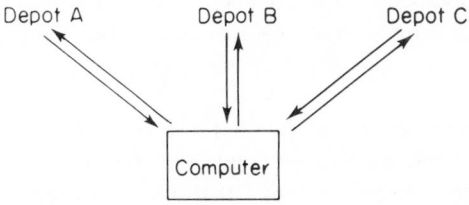

Figure 12.2   A decentralized data input system

The depot personnel in this system experienced difficulties and frustrations with the computer but they did not regard it as a controlling influence in their lives. Rather, they saw it as a tool to help them conduct the daily cycle of activities in the depot and hence to serve their customers better. They also found the teleprocessing system a useful source of contact with the rest of the organization, giving them a sense of identity with the other depots and with the organization as a whole.

Data input options

It is clear that these first two companies have adopted different data capture philosophies. The philosophy of the designers of the first, the centralized system appears to be the classical one of functional specialization, i.e. there is a function to be performed (e.g. data input), therefore dedicate a group of people to this function. The result, in job design terms, is a job which lacks variety and has no perceivable utility, because the operator does not know what the computer does with the information. In addition, the computer can 'pace' and correct the operator, and this leaves him with little autonomy over the pace and quantity of his work. This is the analogue of the paced assembly-line job for the white-collar worker.

It is doubtful whether the designers of the second system had job design criteria in mind, but they created jobs characterized by variety, utility, a sense of cycle completeness, and a strong identification with the work of the group. In particular, the operators provided information in order to obtain output which had to be accurate if the depot was to do its work. This had profound psychological ramifications; whereas the computer in the first case is the unforgiving master, in the second it is a tool from which a good service is required.

There are no technical reasons why the design philosophy adopted in the first case study should not have been adopted in the second, and *vice versa.* From our experience of a number of systems it appears that designers have a considerable range of strategies by which they can capture data and which have different job design consequences. The following guiding principles may help to obtain accurate data capture and at the same time preserve and create jobs which are more psychologically meaningful.

1. *Collect data at source.* If the data are collected as near to their source as possible, the data input load may be spread widely and thinly across a great many people. On the basis of short-term costing this may not be as efficient as the rigid division of labour, but social benefits will more than balance this cost.

2. *Make it part of the data generating transaction.* If possible, make the act of capturing data a part of the process by which the data are generated. For example, in banking it is possible for details of withdrawals or credits to be recorded in computer-compatible form at the same time as the clerk is dealing with the customer. In this way data capture becomes an incidental part of the process.

3. *Ensure each person deals with output as well as input.* Probably the single most important principle is that the computer system should serve some function for the person providing data. Thus the terminal should involve a two-way communication and the person should have tasks to perform which require access to the computer's data base, as in dealing with customers' queries. The output requirement appears to serve the very important psychological function of enabling the person to see the computer as a servant rather than a master. This has been recognized by the Steel Workers' Union in Norway, who have banned the use of one-way data capture terminals.

4. *Input and output data should have a perceivable relation.* As a corollary to the previous principle, it is important that the operator can relate what he enters to the output he receives. This does not mean it must be the same information although that is desirable. It

does mean the operator should know and understand the relationship. For example, the bank cashier may enter details of one customer but may need output for a customer whose details he did not enter. The act of requiring information about a customer gives meaning to the act of entering the information and makes clear the consequences of poor data entry. If this relationship is not clear, data entry can be meaningless 'feeding the machine' and the quality of input may be regarded as unimportant.

5. *Collect the data once only.* There are few things more frustrating to a person than having to undertake an apparently meaningless and unnecessary task. The act of transferring data from one computer system to another computer system can seem such a task with its attendant risk of errors. Integration of computer systems so that the data need only be entered once is not only more efficient but removes the need for the performance of tasks of this kind.

6. *Where data are centrally collected, provide other system-related functions.* It would be foolish to ignore the possibility that in some situations it would not be possible to decentralize data input, and some thought needs to be given to design of jobs in centralized data input functions. In our experience, it is rarely necessary to centralize all data input but it is often necessary to centralize some of it. In these circumstances we have been struck by the variation of responses of operators from one system to another. In some systems operators are apathetic and alienated, but in others they have a deep sense of commitment to the system and enjoyment of their work. The difference appears to derive from the fact that the latter group are fully identified with the system, have a responsibility for the integrity of the data in it, and serve as experts on it to potential users. In other words, these operators are more than data input clerks; they are closely involved with the quality of service the system provides for its users. If data entry clerks are not permitted an opportunity to take these responsibilities, they are mere appendages of the machine and will respond as such.

The example of data input illustrates the need for the design process to take explicit account of job design as well as technical considerations. In practice there may be many issues involved in the creation of jobs other than the distribution of data input. To illustrate some further issues and to consider more closely the process of computer systems design we will now examine another case study, this time a longitudinal study of the introduction and development of a computer-based on-line order processing system into the commercial head office of a large electronics company.

The introduction of a computer-based on-line order system

The company

This company markets products from six United Kingdom factories and from other associated factories throughout the world and sells to large users of electronic components distributors in the United Kingdom, and associated companies in other countries. It is operating in an area of rapidly changing technology with a change in product base because of the development of new products such as integrated circuits and microprocessors and the continuing obsolescence of many of its previously long-running and profitable products.

Because of the nature of its business, there is a complex relationship between manufacturing and marketing. The total range of components covers 15,000 different items grouped into 16 product groups varying from small resistors to TV picture tubes. These products are supplied in different combinations across the whole area of industry.

This complex market/manufacturing relationship means that there is no one ideal way of organizing the business, so there is a history of continual reorganization. At different times over the past 10 years the organization has been built up on either a market or a product base with tendencies towards a matrix organization without a wholehearted commitment to it. At present it is basically market oriented, but recent changes have moved it slightly towards a more product-based design.

The changes outlined here are taking place in a changing employment environment. The total staff employed in the company's head office has dropped over the 10 years 1969–1979 from around 1200 to 700, largely by natural wastage but with some redundancy.

It does not yet have a high degree of union membership although, as with most other clerical operations, there is a continuing increase in the proportion of the staff who are union members. The total organization of which it is a part has a positive attitude towards the involvement of unions in the traditional areas of negotiation and it might be expected that union membership would have happened earlier here, but many staff at all levels have apparently seen no advantage to themselves in joining.

History of computer systems

The first moves towards computer-based order handling were made in 1964, with a system which computerized all stock control and ordering activites working in a batch processing mode.

In 1972 a more sophisticated change took place, combining visual display units and computer-prepared order cards for each line on an order, with data input done at the order desk rather than the computer. The introduction of this system reduced the amount of paper flow and caused many simple tasks to be redesigned or to be removed. The management felt that many people were not capable of the comparatively higher-level jobs (though still actually at a low clerical level) now required and many of the lower-level people left. Job rotation was introduced between many of the order entry and order preparation tasks to give some variety to the jobs.

One of the major problems with these systems was the impossibility of maintaining on a day-by-day basis agreement between paper stock and actual stock, leading to competition for anticipated stock and the possibility that some items could be committed to different customers by order clerks not knowing of other clerks' activities. The situation was complicated by the system being relatively slow to operate in the proper way, so there was a lot of bypassing of it by urgent orders which increased the risk of stock discrepancy.

Description of the on-line ordering system

The on-line real-time order book (OLOB) system which commenced operation in mid-1976 is one of the largest real-time systems in the United Kingdom. Orders are received for the 15,000 different items (including all variations, standards, etc.) from 6000 customers at a rate of 180 orders per day, each with an average of three lines. There are 95 visual display units in use and the system is working with an average daily usage at present of 25,000 messages per day, which is still increasing.

Each order is placed on the computer by an order clerk using a VDU; the stock situation for each item is interrogated *via* the computer and the delivery schedule plan. The customer is offered alternatives if his order cannot be satisfied as required and the order is processed into the computer system. All real stock information is updated as each order committing delivery of an item automatically changes the stock indication for that item. Orders for future delivery are recorded. Up-to-date information can be obtained on each order, the order book for any one customer, stock, or each item in various levels of detail.

Within this system framework it is possible to conceive of a number of ways in which work at the order desks could be organized. One way would be to create a *functional* division similar to that found in the centralized data input system discussed above. Thus a group of clerks could be solely concerned with entering orders *via* VDUs which have previously been finalized by other clerks who deal with the customers. Alternatively, the division of work may be by *product* (or group of products) or by *customer* (or group of customers), each clerk carrying out the full range of functions relating to the product or customer. A combination is also possible in which clerks may be responsible for a limited range of products for a group of customers.

From the point of view of job design it is better if they each individually, or as a group, have a pattern of tasks which they see as making up the complete job for which they feel responsible. Product or customer groupings meet this principle, and the principles enumerated above in respect of data input, much better than the functional division.

It is clear that these solutions are difficult because clerks must interlock with others in the system. If they are dealing with a group of customers they cannot have total control over an individual product (because other customers will be buying it); if they have responsibility for a particular product it is difficult for them to see the total needs of a customer and it makes it more complicated for a customer in that he has to talk to a number of different people to deal with one order.

The present solution adopted by the company is that the orders for any particular *customer* will come into a group of clerks who will deal with it on a *product* basis, i.e it would pass from one to another who would deal with each particular product group. There are differences between the various divisions as to how many of the functions, editing, pricing, entering, commitment of delivery, and progress chasing, each individual will carry out, and these differences reflect the previous ways of organizing work in the various divisions.

The computer system brings together all the orders for a particular product so that the product planner can obtain supplies from the appropriate UK factory or overseas company.

The Order Office is notified of the fact that goods are being released from stock for delivery to the customer by two methods, both involving the use of hard copy. If stock is available for all orders it is released automatically and the Order Office notified that delivery is to take place; if committed orders for an item exceed the stock available because the factory has not fulfilled its promised production, the Order Office has to decide how the stock should be allocated. The Order Office can also authorize despatch without being notified by the computer.

The process of introduction of the OLOB system

Because this company is in the electronics business, it has a history of computer usage which makes it a relatively sophisticated user compared with many other companies. The main process used for considering the introduction of OLOB was a number of interlocking committees involving people from within the head office and from an internal computer consulting organization.

In general, the Order Office management were fairly well involved in discussions concerned with developing the concept of the system. The clerical staff were involved to some extent in all divisions but the degree varied from one to another. Generally, the clerical staff were kept informed by informal departmental meetings at which they were given proposals and invited to comment. These sessions were largely seen as a way of 'selling' the management proposals to the staff rather than as a means of getting involvement of the staff in participating in the design of the system.

To some extent, the differences in approach again reflect the differing degrees of commitment the various divisions had to the idea of computers based on their previous differing experiences. The sales managers and their engineers who would have to work with but not actually use the system were not much involved in its introduction.

OLOB required keyboard skills from a wider range of people including those in higher level clerical and low to middle level management jobs and some resistance was expected and encountered because of this. Training was given at different levels according to the expected degree of usage and the objections have now all disappeared with familiarity and proficiency.

Technical and efficiency effects

From an efficiency point of view the OLOB system is considered a success. There is now higher turnover of goods, a lower level of stocks, and fewer staff are required to operate the system.

Better information is now available on the value of the forward orders and on the degree to which promises to customers are not being met. There is now less discrepancy between actual and computed stock levels and the need to bypass the system has been reduced to a very low level.

There are still a number of dissatisfactions with the system some of which were foreseen but could not be incorporated and others which were not foreseen because of the difficulty in understanding the effects of an unknown system. The introduction of a computer system can reduce flexibility and unless the need for an operation has been recognized in advance it becomes more difficult to carry out.

Effects on jobs

The overall feeling of those working the system is that they would not want to return to the older way of working with its many frustrations. The overall level of job satisfaction is now much higher than previously. The major concern, as would be expected, has been with the number of jobs likely to be lost but this has really been quite small with the introduction of OLOB as compared with previous changes. The OLOB system as installed is very flexible in that it allows many choices as to how the various tasks are packaged into jobs. An individual can perform one function or several tasks can be built up into more complete jobs.

Each division has moved towards a philosophy of having more complete jobs. The way in which this has happened depended on the previous ways of organizing work. The division which had the least history of computer involvement started up with each individual performing only one function but set up in parallel a pilot group where some people did the whole range of functions in relation to the customer and others similarly in relation to the product. It has now moved to adopt the latter as the normal way of working.

Some managers felt that there had previously been a policy of recruiting lower-level staff who were only suitable for routine clerical operations, and on one order desk about a third of the staff have not been able to adapt to the higher-level jobs which have been introduced and the changes have taken longer. In this division staff do have some choice as to whether they progress to higher-level jobs or not, and as new staff are recruited they are told that they are expected to progress to being able to carry out the whole job and their pay is increased as they become more versatile.

Some managers claim that older people in low-level jobs have left because they could not cope with the change but the evidence for this is scanty. There is one older man working happily and efficiently with his VDU who was previously employed on low-level administrative duties in the garage.

In general, the sales order organization is now flatter, with staff of a higher average grading doing a higher-level range of work. Some clerical staff have taken over the roles of office-based sales engineers and are able to offer alternative products if the type ordered is not available. The number of inside sales engineer post consequently has reduced.

A major factor in determining the quality of working life appears to be contact with the world outside, dealing direct by telephone, etc., with customers or with factories.

Job design in the design and implementation of computer systems

The OLOB case study resulted in reasonably successful job design and it is instructive to ask how it was achieved. Although the jobs developed have many of the features found to be important by job design research they were not created in the light of expert knowledge of this research. Two factors appear to have been important in the formation of these jobs:

(a)   The participation of potential users (or potential user management) in the design process, ensuring consideration of issues of importance to users and the perpetuation of traditional user procedures;

(b)   A flexibility in the design of the system which enabled each user department to *evolve* a way of creating jobs linked to the system which met their needs. Thus the system supports several different job structures simultaneously.

Of these factors the question of user involvement in systems design has been attracting considerable attention. Indeed, in Norway there is now a legal requirement that user representatives be involved in systems design (Docherty, 1980).

The variety of methods used to provide user involvement is outlined in Eason and Damodaran (1979). It ranges from communication through newsletters to user representatives joining steering committees or becoming full-time members of the design team. The evidence suggests that none of these techniques goes far enough to enable all potential end-users to play a part in the construction of their future jobs. As a result more radical practices are now being adopted under the heading of Participatory Job Design. This approach is at its most advanced in the Scandinavian countries, and case study reports of its use in computer systems design are beginning to appear (Hedberg, 1980; Clausen, 1978). In the UK, Mumford (1980) has reported a number of similar case studies.

The general principle of this form of design is that end-users should be involved in selecting the objectives of the system and the form of work organization they themselves will operate, i.e. the tasks they will undertake, the manner in which tasks are combined to form jobs, the relationships that exist between jobs, and the relationships that would exist between staff and the computer system. The end-users are not expected to program the system or to engage in technical aspects of system design, these functions remaining with the technical expert.

Mumford reports that when this approach works it leads to a successful socio-technical system which is adopted with enthusiasm and commitment by its users. However, it is a difficult course to follow and is frequently unsuccessful. The main problems appear to be as follows:

1. *The lack of knowledge of the users.* To appreciate their options and constructively choose between them users need knowledge about the flexibility of computer systems and the kind of impact particular forms of system may be expected to have. In particular, they have to be able to assess the implications of particular designs for their own jobs and this is usually very difficult for them. Most projects which adopt participative job design include resource personnel who attempt to provide this knowledge as and when required and to do so in a form which can be readily understood.
2. *The role of the technical expert.* In this approach the technical expert takes on the role of servant or adviser to the user group. This is an

unfamiliar and possibly an uncomfortable role for the computer specialist, and Hedberg (1980) points to a case in which the users formulated a system specification but the technical experts, in undertaking the technical design, radically changed the nature of the specification.

3. *The size of the system.* Successful reports of computer systems design employing participative methods have usually involved small-scale systems for a small number of users whose tasks were relatively interdependent and somewhat separate from the tasks of others. Where a large-scale system and a large population are involved participative design can be unwieldy. The case reported by Hedberg (1980) is an example. Attempts to involve all of the users in a building society in the design of a computer system worked well in the initial systems specification but, as the detail began to be worked out, the process gradually became dominated by systems staff and management and the planned work organization gradually reverted to the *status quo.*

4. *The degree of participation.* The aim of participation is to involve everyone who will be working with the new system but this is difficult to achieve, especially in large organizations. Some form of representative procedure is usually employed, especially if trade unions are recognized, but the representatives do not necessarily discuss with their constituents the proposals being considered nor do they put forward their views. Steering groups similarly do not usually involve all those concerned with the operation of the system. They may miss out the lower levels, as in the OLOB case study, or omit user groups who are not the prime users. There are two main difficulties which arise in any participative approach, the concern about job loss and the need of some people to preserve their own vested interests, and these can cause managements to doubt the wisdom of trying to please everybody.

The problems attendant upon the use of participative methods cause us to doubt whether it can be a sufficient strategy alone to ensure the best job design outcome. Fundamentally we question whether even the users themselves can take the best once-and-for-all decisions about the kind of jobs they should have within the tight constraints and deadlines of the normal systems design process. Users need time to understand the consequences and opportunities of the new technology and this demands flexibility, with the system permitting an evolution of ideas amongst users. There is an additional need for flexibility beyond the implementation stage, because users will continue to learn and organizational change will continue to be necessary. It is our view, therefore, that the flexibility for evolution of job patterns found in the OLOB case study is a very important factor if job design is to be successfully achieved.

A case study in evolutionary design

The final case study we wish to consider is one in which the twin aims of participatory design and evolutionary design are being sought. At the time of writing the system is in the process of being designed. It will be a major on-line, real-time system to support documentation handling and interbranch coordination in a freight-forwarding organization which operates 30 branches throughout the UK. The strategy being adopted is one in which the major emphasis is upon the education of staff at all levels in order that they may take informed decisions about the computer systems they need; this strategy has two main features.

One feature is that an evolutionary pattern of computer introduction will be followed. Thus a computer system is being introduced as a pilot scheme to one branch in order that this branch and the rest of the organization may learn from practical experience the implications of computer use in their business. It may be that the result of their study is the discontinuation of this particular system and its replacement with an entirely different system, although this may be difficult once the idea has gained momentum. Thus the pilot scheme is essentially regarded as a realistic simulation which would enable potential users to react to real terminals, procedures, printouts, etc., rather than abstract flow charts.

This proposal is given additional weight by looking at the experience of the OLOB system described above. Effort was expended in trying to involve the users in the design of the system but some problems only became apparent when they had the first VDUs installed in the working environment and they could try them out themselves. In the OLOB case the system was designed to be flexible so that there was a lot of freedom to decide how it should actually work in practice. An important aspect of the rapid development of computer technology, however, is that it can now be cost effective to consider simulation exercises before embarking upon major computerization programmes.

The second feature is that the education of users should be planned to be progressive and to fit the rate at which computer usage within the organization is evolving. Thus a series of seminars was undertaken for the directors and other senior members of staff in the freight-forwarding organization to familiarize them with computer developments in general and specifically in the context of their business. This series of seminars ended with sessions on the management of change leading to discussions of the kind of organization sought for the future and the kind of computer systems that would support it. As these discussions become more specific, so evidence from the pilot study will be brought into consideration and the participation process spread to other levels and other branches of the organization.

It is too early to know whether this process will be successful. It does make the two points, however, that, when large-scale computerization is planned, it is important that the participative process begins at the top of the organization (and that the need for participation at all levels is emphasized) and that concrete and specific data about the impact of computer systems are generated as early as possible and fed into the participatory process.

Conclusions

It is inevitable that the introduction of information processing technology should have an effect upon those jobs which consist largely of information processing tasks. The case studies presented in this chapter clearly indicate that there are many options in the way this technology influences jobs and that, by consideration of these options in the design process, successful job design may be achieved. It is, however, by no means a simple task to consider job design options within the system design process. It is particularly important that technical system designers should not make judgements about 'good' jobs for other people but that the future job occupants should participate in the design process. This in turn raises questions about the process of effective participation and it is our conclusion that users require an evolutionary systems design framework in order to gain the experience necessary to take the wisest decisions on their own behalf.

References

Clausen, H. (1978). Concepts and experiences with participative design approaches. Proceedings of the Conference on 'Design and Implementation of Computer Based Planning Systems' (Ed. N. Szyperski), Cologne, September.

Docherty, P. (1980). Some consequences of Acts and Agreements in the Scandinavian countries regulating user participation and influence in system design. In 'The human side of information processing' (Ed. Bjorn-Andersen, N.). North-Holland, Amsterdam.

Eason, K. D., and Damodaran, L. (1979). Design procedures for user involvement and user support. Infotech State of the Art Report on 'Man–Computer Communication' (Ed. B. Shackel).

Hedberg, B. (1980). Using computerised information systems to design better organisations and jobs. In 'The human side of information processing' (Ed. Bjorn-Andersen, N.). North-Holland, Amsterdam.

Mumford, E. (1980). Participative design of clerical information systems. In 'The human side of information processing' (Ed. Bjorn-Andersen, N.). North-Holland, Amsterdam.

Stress, Work Design, and Productivity
Edited by E. N. Corlett and J. Richardson
© 1981 John Wiley & Sons Ltd

Chapter 13

An Ergonomics Study of a Press Workshop with the Objective of Improving Working Conditions in a New Factory

F. Jankovsky
Laboratoire de Physiologie du Travail et d'Ergonomie,
Conservatoire National des Arts et Metiers, Paris, France

Introduction

The management of a large light engineering firm had decided to build new workshops as there was a severe lack of space in their present production facilities. The workshops occupied premises which were previously used for manufacturing textiles and the buildings had not been designed to hold a machine shop. In addition, as the firm developed, all the available space was gradually used up and it had become apparent that saturation level had been reached; there was no more room to install new workplaces and the general overcrowding had begun to hinder movement around the factory.

As the firm wished to continue expanding it decided that it was necessary to build new factory premises. The firm had decided to commission a study on the working conditions in their present factory. The study had two main objectives:

1. To put forward recommendations on how to provide better working conditions in the new factory.
2. To present a number of propositions concerning 'task enrichment' for the machine operators.

The project proposals were originally submitted to the Agence Nationale pour l'Amelioration des Conditions du Travail, who in turn presented them to the Laboratoire de Physiologie du Travail et d'Ergonomie du CNAM.

The study was organized into four distinct phases which form the basis of a methodological approach for industrial ergonomics studies: analysis of the project proposal, analysis of the work carried out by the operators, presentation of results, evaluation of the project's impact.

This chapter has attempted to highlight the methodology employed in the study, as one of the project's principal objectives concerned its development.

Analysis of the ergonomics project proposal

This phase was set in motion with dicussions between a representative from Agence Nationale pour l'Amelioration des Conditions du Travail (ANACT), the Director of the Laboratoire, and the members of the team who would be carrying out the study.

The meetings with ANACT provided background information on the firm, its situation, and how the study fitted in with the construction programme. The two specific points in the proposal were discussed in more detail, i.e. recommendations on how to improve the working conditions in the new workshops and proposals on how the operator's tasks could be enriched.

Regarding the first point, the study team explained what sort of contribution to improving working conditions could be expected from an ergonomics study. However, there were a number of reservations concerning the second point. Task enrichment presents a complex problem which could not easily be resolved within the time limits involved in the construction programme.

After the Laboratoire de Physiologie du Travail et d'Ergonomie du CNAM had finally agreed to undertake the project, the research team organized a visit to the firm. The researchers had asked to meet the factory director, the employee's representatives, the union representatives, and the medical health officer. They also asked to visit the workshops and to have the opportunity of meeting with the departmental managers.

This first phase of the study was carried out with two principal objectives in mind:

1. To gather sufficient information to define both the study goals and the firm's reasons for initiating the original project proposal.
2. To provide a means of introducing the research team to the persons concerned with the study in the firm and of informing them about the content of the study, its programme, the methods employed, and the way in which the results were to be presented.

The first phase was completed by two members of the research team on a two-day visit to the firm. A summary of the information they collected is given in the next section and is followed by a description of how the research team briefed the firm on the study.

Information collected

The firm is a subsidiary of a large French industrial group which produces household consumer goods. The factory where the study was carried out belongs to a regional group consisting of four manufacturing units which produce a range of one type of product. Three units manufacture components which are later assembled in the fourth factory.

During the visit to the assembly plant, it was observed that the production organization was highly dependent on the quality of the components coming from the other three units. The components were required to conform to very strict tolerance limits as there were no facilities at the assembly plant for rectifying individual components which presented problems in the assembly operations. Thus, quality standards were the determining factor in defining the task demands in the component manufacturing units. The quality control requirements were strict for both engineering precision and visual appearance. The firm's marketing policy is to release onto the market products of only the highest quality.

At the meeting with the factory manager, the plans of the new factory, which had already been drawn up, were shown to the study team. The new site had been chosen and was not far from the present factory. It consisted of a narrow strip of land which determined both the width and the general orientation of the building. As it was not possible to use another site, the only way of increasing the overall surface area of the factory was to increase the length. During the discussions the factory manager requested that the report on the ergonomics study should be presented to the firm within one month of the observation phase so as not to delay the construction programme.

The visit to the press workshop impressed on the study team just how serious the overcrowding problem had become: the workplaces were cramped, machines were situated very close to each other, trolley corridors were very narrow, raw materials and machined parts were piled around (see Figure 13.1).

The workshop consisted of 40 presses, the majority of which were operated by women. The presses could be divided into three classes: single-stroke presses, semi-automatic and automatic. A number of the workplaces were organized into a 'production line' where the product moved from one press to the next for successive operations and each press was worked by one operator.

The high noise levels in the workshop were immediately noticeable, rendering normal conversation very difficult. Subsequently, the medical

Figure 13.1   Photograph of the press workshop

health officer remarked that he had diagnosed the onset of hearing loss in certain workers.

A meeting had also been arranged with the managers in charge of the different departments in the factory (production, maintenance, methods study, planning, quality control). This meeting provided further background information necessary to fully understand the work carried out on the shop floor and it particularly highlighted a number of important points:

—It was confirmed that quality control standards played an important role in the production process; occasionally whole lots were returned to the factory by the assembly plant. Certain training problems existed: for example, 'Every morning I have to teach the operators how to carry out inspection controls' said the manager.

—The tasks were more complex than was immediately apparent: 'There are good and bad workers: with the good workers, tools on the press last a long time, with the bad workers the tools have to be changed frequently, and sometimes they are even cracked, which can cause the machine to be stopped for quite a few days' said the maintenance manager.

—The shop-floor representatives pointed out that more acute problems of working conditions existed in workshops other than the press shops (cleaning and polishing shops). They proposed that the research team should also study these areas and put forward suggestions to improve working conditions there.

—The managing director stated that the firm was still expanding from an economic point of view. This constituted one of the main reasons for building a new factory, as the lack of space in the present facilities restricted the opportunity for any development.

—However, even taking into consideration the favourable economic situation, the firm was still obliged to employ temporary workers. The managing director was trying to avoid employing this category of worker and to provide permanent jobs. However, a number of different constraints had prevented him from being able to offer more stable contracts.

Exchange of information between the study team and the firm's employees

During the initial contacts made in the meetings described above the research team found it necessary to define the scope and limits of ergonomics and to explain how an ergonomics approach would be applied in this study.

With this in mind, the project team introduced itself as a group of ergonomics practitioners and clearly defined the field of study as being centred around the task carried out by the operators on the shop floor and the different factors—economic, technological and organizational—which governed or influenced the task. This approach also required that certain information on the working population be collected: sex, age, experience, training courses followed, length of training periods, and so forth.

The project team also explained its position as a group of research workers based in a laboratory which is funded by public money and whose function is to carry out various types of research and to teach courses related to real problems in industry. Industry benefits from these studies by being able to exploit the results to resolve specific problems.

It was also pointed out that the recommendations would only concern the requests made in the project proposals, namely, how to improve working conditions in the new workshops compared to those which existed in the factory where the study was undertaken. It was also opportune to add that this approach was limited by two factors—first that the field of ergonomics far from covers all the different components presented by work in the factory, and second that any decisions concerning the implementation of the project recommendations were outside the study team's terms of reference, as it did not have any executive power within the firm.

The methods which were to be employed during the running of the project were also given to the interested parties in the factory. It was necessary that an observer be present in the workshop to carry out observations for a task analysis. In addition to the observation techniques employed the operators had to be interviewed on the different aspects of their work. Similarly, certain equipment was required to gather information on the work activities carried out and on the work environment: event recorders, film or video, tape recorders, sound-level meters, dose meters, photometers, thermometers, etc.

Any measurements or recording had to be carried out as unobtrusively as possible so as not to disturb the operator in the execution of his task. In any case, it was necessary to obtain the operator's consent before any measurements were carried out; the aim of the project and the techniques used were also to be explained.

Finally, when discussing the methods to be employed, the way the results and conclusions were to be presented had to be determined. Here the firm was told that a report would be written after the field study had been carried out. It was suggested that the management should circulate the report to the union officers and the shop-floor representatives. The study team would then return to the factory after each group had had the chance to read the report to explain any points and to discuss how any recommendations on workplace layout could be implemented.

At the end of this first visit to the factory, during a general meeting with the representatives of all the three groups (managers, unions, workers), a number of specific points were raised:

—   The study would only consider the press shops, the other workshops were not within the project brief.

— The field study would involve four ergonomists visiting the site for one week at a prearranged date.
— A report would be completed three weeks after the data collection in the factory. The report would be circulated to the different groups represented in the firm.
— The results from the study would be used as the basis for recommendations:
   on certain specific characteristics of the architectural layout;
   concerning the interior layout of the workshop;
   specifying the type and number of presses which would eventually be transferred to the new factory.
   concerning the installation and equipping of the individual workplaces;
   on task organization and content.

Task analysis on the presses

This phase was carried out in five days by four members of the team, according to the programme decided on in the initial meetings.

The week started with a meeting for all the personnel in the workshop for an introductory talk. The objectives and reasons for the study team's presence in the workshop were explained to the workers. This was necessary because the press operators had not been previously informed that a new workshop was planned and that a study was to be carried out beforehand.

It should be noted at this point that the study team had not met the press shop workers as a group during the preceding phase, only their representatives; on any field study it should be ensured that the people involved are informed about the project before any observations are made at the workplace. In the present study the shop floor should have been given the same information as the other employee group representatives from the beginning.

A task analysis was carried out on each press which was working during the data collection phase. The presses were classified into different categories of workplace type. The classes were based on the technological characteristics of the machine—manual presses, semi-automatic presses, and automatic presses. The manual press category was divided, using an organizational criterion, into individual press workplaces or presses grouped into a production line.

The press workplaces could have been classified using other criteria, for example by press operator category (temporary staff, skill class 1, skill class 2) or by the type of component being produced. However, it is important that stable criteria be chosen so that the results can be generalized and exploited in the best possible way. The type of component produced at the press does not provide a useful criterion for classifying the workplace, mainly because of the variety of components produced, consequently each press produces a range. Similarly, operators rotated from workplace to workplace, so operator category was not acceptable as a useful criterion.

The discussion here will be limited to results taken from the task analysis for manual and automatic presses.

Manual presses

On this type of press, it is immediately noticeable that the task organization is based on repetitive operations.

During the interview the operators remarked that there existed production goals, laid down by the work study department, which were used to calculate an overall bonus for the workshop amounting to approximately 5% of their salary.

The operators were asked how many items they had to produce per shift. The results are given in Figure 13.2, with the mean theoretical cycle time $(\bar{M}_T)$ calculated by dividing total work time per shift (480 mins) by the daily production norm in number of components.

Press	Daily production levels Number of pieces produced/day (N)	Theoretical cycle time (sec) $\bar{M}_T = \dfrac{480 \text{ min}}{N}$	Remarks made by operators
No. 1	2800	10.3	The same production level is maintained for painted sheet, although in this case it is necessary to carry out an additional greasing operation on each item.
No. 2	4500	6.4	At rate 85.
No. 3	3200	9.0	
No. 4	6000	4.8	Very difficult to achieve. It would be acceptable norm for 4500–5000 items per day.
No. 5	2800	10.3	At rate 100.
No. 6	4000	7.2	

Figure 13.2   The rates 85 and 100 refer to work rates defined by the methods department; the maximum bonus is obtained for output corresponding to a production rate of 100

The task was observed over a number of cycles using a systematic technique for recording 'critical incidents'. The results are presented by histograms in Figures 13.3 and 13.4. They show that 'critical incidents' are responsible for the large variation in cycle times compared to the theoretical mean. In the example given in Figure 13.3 the norm (rate 100) is 4000 components a shift, which gives a mean cycle time ($\bar{M}_T$) equal to 7.2 secs.

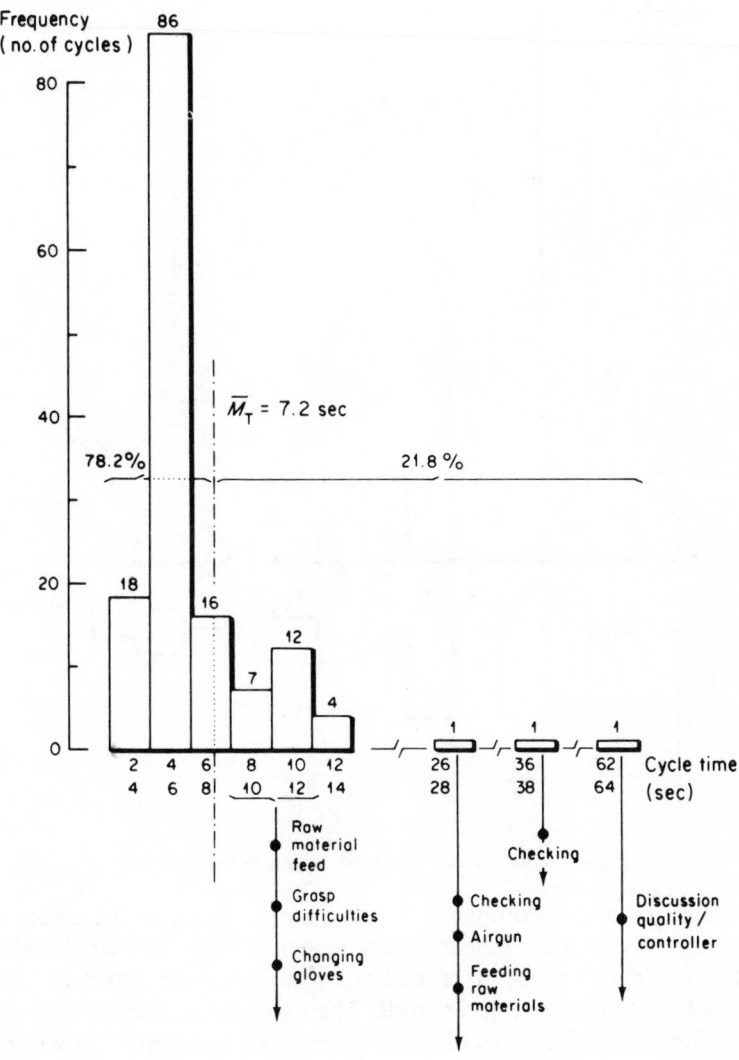

Figure 13.3 Critical incidents on one press

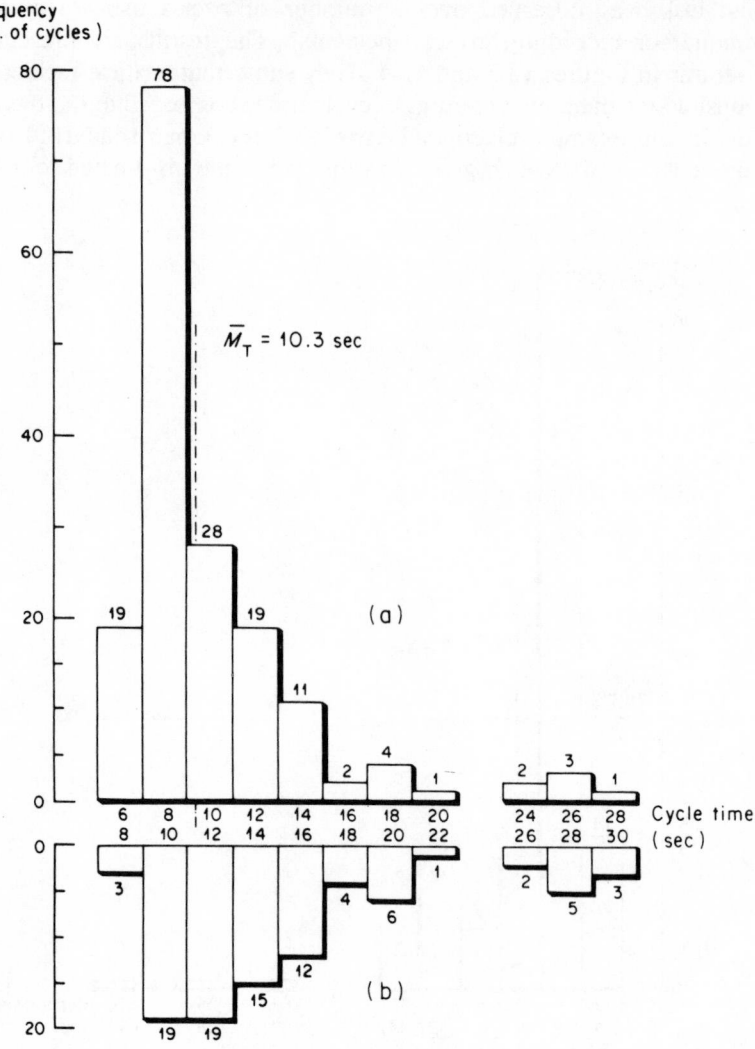

Figure 13.4   Critical incidents during press operation

The number of cycles completed in less than or equal to the mean $\bar{M}_T$ is high, 78.2%. Cycle times longer than the mean, 22.8%, corresponded either to additional operations being carried out (feeding in raw materials, clearing away waste) or to inspection controls. These subtasks necessarily take time and are inherent in the production of 'good components'; however, the operator has to complete other components in times shorter than the mean cycle time if production goals are to be maintained. There is a limit to the cycle time which can be attained; it is impossible to make a component in less than 2 secs. Consequently, a work cycle of 62–64 secs (see Figure 13.3) corresponds to a theoretical production time for eight components if $\bar{M}_T$ is taken as a reference. In order to catch up the lost time of about 55 secs (62 minus 7,2 = 55), the operator has to produce 14 components at the fastest rate possible or 27 components at the modal time (cycle times of between 4 and 6 secs).

A series of systematic observations was made at another press during 168 successive cycles. The 'critical incidents' occurred also during the shorter cycles. The results are shown in Figure 13.4.

The 'critical incidents' have been classified as follows:

Extended inspection	6%
Adjusting/greasing	4%
Difficulty in grasping a component	5%
Difficulty in positioning the component on the press	14%
Difficulty in extracting the component or waste from the press	71%

These formed the basis of a task evaluation from which certain ergonomics proposals for modifying both the task design and the workplace layout were derived.

A similar series of observations was made on each press in the production line group. The 'critical incidents' fell into the same categories as those in the previous case but additional types of incidents were noted which originated from the organizational dependence of one workplace on its neighbour. The operators on the line were even more constrained and had less control over their task activities, because of the production organization. They not only dealt with the incidents which arose on their own press but also felt the effect of incidents which occurred on posts above and below their workplace on the line.

Each operator's postures were also recorded during the periods of systematic observations of critical incidents with the view of evaluating the influence of the task on the type of posture operators adopted.

To conclude, work on the manual presses can be characterized by the speed and repetitiveness of the operations. The operators had to fight against the time constraints if they were to produce the number of items required to meet the production goals and ensure that quality standards were respected and the machines maintained in good order. The operators worked for over half the shift time (between 57.7 and 80.5% of the observation periods) at a rate faster than that defined by the work study department.

The repetitiveness of the task cycle resulted in specific constraints on posture. The operators worked seated with their legs in one position determined by the seat and press construction. The head, arms, and trunk were continually in motion but the same muscle groups were exercised from cycle to cycle. When the operators had produced 2400–6000 components a day, they had made 2400–6000 times the same or very similar movements.

These characteristics, in conjunction with the speed and repetitiveness of the operations, were also relevant when considering the mental load of the operators. Constant visual attention was a requirement necessary to predict the occurrence of certain critical incidents and to initiate corrective actions in response to random events in time.

Automatic presses

According to the technical specifications the automatic presses ought to produce 10,000 components per hour. In fact, the daily production levels were estimated to be between 20,000 and 30,000 items per day.

The interviews with the operators provided two important points of information. The work from the automatic presses did not count towards the workshop's productivity bonus and the workers on these presses belonged in general to higher qualified categories (OS2: semi-skilled class II).

It was apparent from the first set of observations that the presses worked in a large number of periods of machine running and machine idle. The times for the alternating states of running and stopped were recorded and are presented in Figure 13.5. The mean working time was 110 sec and the mean idle time 32 sec. The longest continuous working period was 153 sec and idle periods could reach 51 sec.

The machine had to be stopped frequently so the operator could clear away waste stuck to the die and avoid damaging the tool and components. When the press was running, the operator's time was taken up by inspection and control of the press itself and the items coming off it, the raw material feed mechanisms, and the lubrication system.

It can be seen from the data that as the running periods decreased the idle periods increased when the steel roll came to an end and during the first few items when a new roll had just been put on. The steel sheet was slightly deformed on the outside and on the inside due to the shocks received during the handling of the rolls. The operators had to pay special attention when the steel was first fed onto the press and at the end of a roll.

The results obtained from systematic observation of the press running suggested the need to study in more detail the control activities exercised by the operators on the automatic presses. The directions where the operators looked during the task were systematically observed on a number of presses. The results from one press, No. 33, are presented in Figures 13.6 and

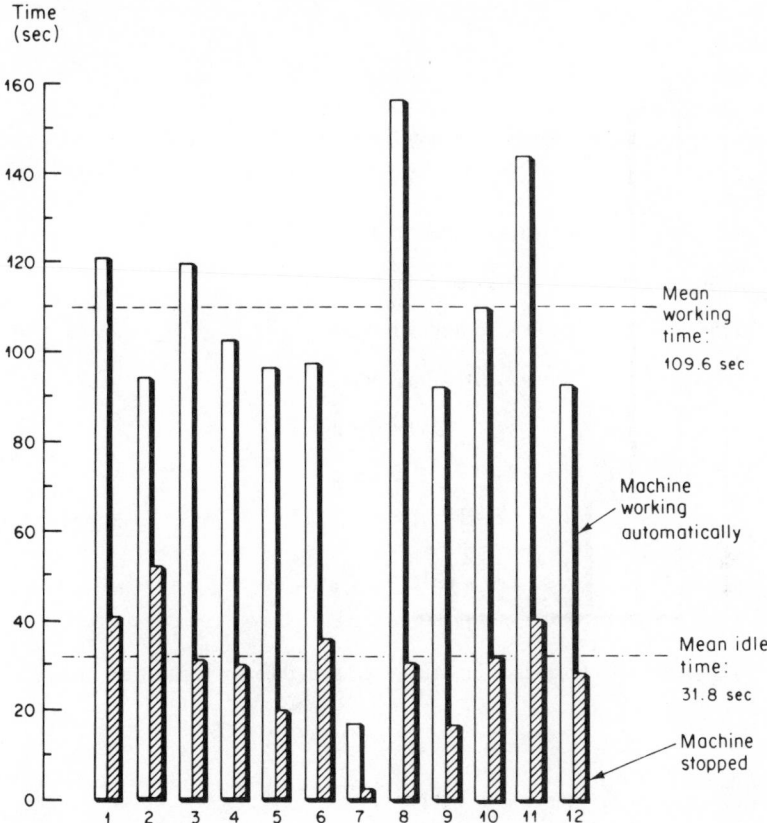

Figure 13.5   Working periods for an automatic press. At the 7th working period, the press was stopped after 17 sec for 3 sec as the operator was talking to the quality controller. During this time the operator seized the opportunity to clear out waste accumulated from the 17 sec working time. This probably explains why the following working period was considerably longer (153 sec). Similarly, the arrival of the other quality controller at the 11th working period explains the duration of this working period. The 7th period was not included in the calculation of the mean working time (109.6 sec) and mean idle time (31.8 sec)

13.7. The operator appeared to control five particular points consistently and other points more randomly (these are classed as 'Elsewhere' in Figure 13.6, which does not necessarily mean that these areas are unimportant). In another category are the controls too brief to be distinguished by observation.

The observations were complemented by a series of interviews with the press operators. One operator explained that he controlled a number of

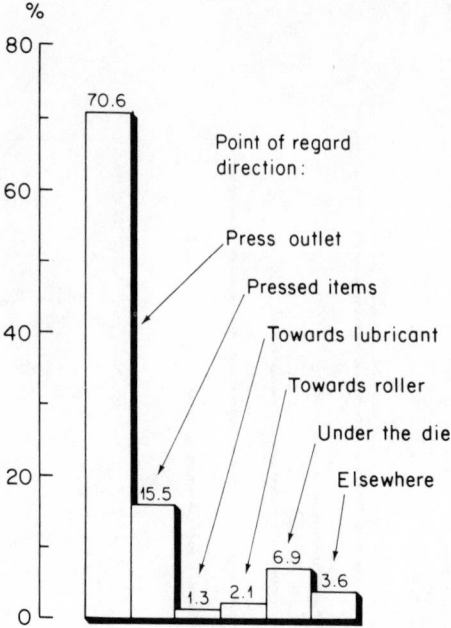

Figure 13.6   Frequency of visual checks on different parts of the press

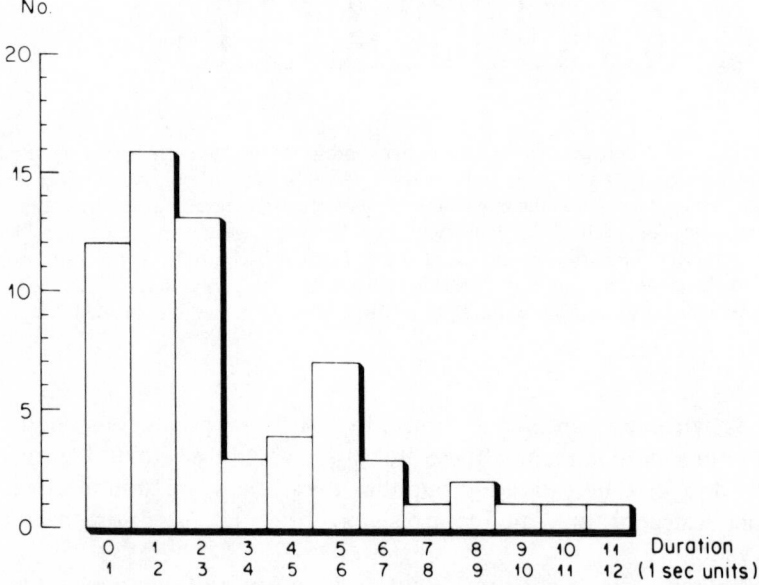

Figure 13.7   Durations of visual checks on different parts of the press

points when he looked towards the item as it left the press. He checked:

— That the component was smooth;
— That the component was clearly cut;
— That the guides were not twisted;
— That the component moved correctly onto the guide;
— That the steel sheet advanced correctly.

The observations and interviews provided a method for defining the task's perceptual requirements and the information processing load of the operator.

The systematic collection of 'point of regard' data also provided a means of estimating how long the operator looked in any one direction (see Figure 13.8). The results showed that there was a considerable difference between

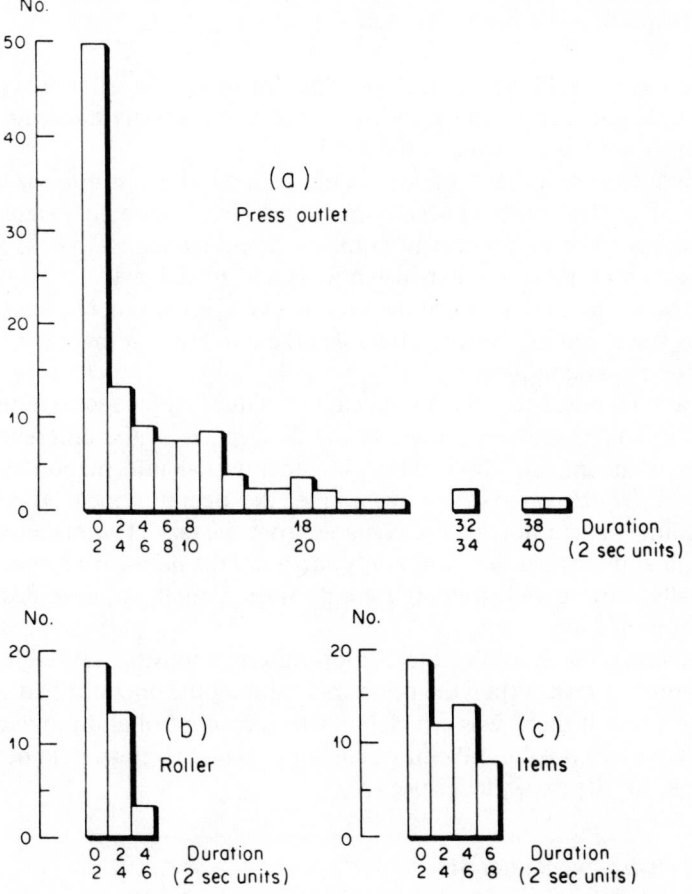

Figure 13.8   Duration of visual attention on different points on the press

the different directions where checks were made and that the range of times varied between particular directions. The times for the checks on the item leaving the press were the most variable, which indicated its importance. The longer durations can be interpreted as the difficulty in identifying information on a particular component or as observing the movements of a continuous process. Checks on the components varied less but were relatively long, while checks on the lubrication system or the feed system were shorter.

The operators' interviews showed that there was a certain overlap between the different sources of information, for example the state of the lubrication system could be checked by:

— Inspecting the glass reservoirs;
— Making sure that the item was amply covered with oil;
— Inspecting the press tool and die.

The control could be carried out directly using the oil reservoirs, but normally the other two indicators were used first and only afterwards was a check made on the oil level in the reservoir.

The work on automatic presses could essentially be summarized into a number of control tasks. These control tasks were directly related to the precision involved in producing complex components as well as the care needed to avoid damaging expensive tools and machinery.

The visual attention aspect of the task was very important (77 % of the task time was spent making various checks) and the interpretation of the available information was complex.

It should be noted that the analytical techniques employed here presented certain limitations and consequently the description of the different control activities is incomplete. In addition to other visual information not taken account of by the observation technique, the operator took advantage of acoustic and tactile information available from the task. The machine speeds were high and the operator could only carry out the necessary inspections by continually starting and stopping the machines, which had been designed to work continuously.

The operators were obliged to compromise in choosing a suitable posture to carry out the task. When the press was running the operator had to have a hand (or a foot if there was a pedal) on the stop control and move the head and eyes to carry out the difficult visual controls and to grasp hold of the item as it came off the press, to inspect it.

Physical environment analysis

After the task analysis was completed a number of environmental parameters were measured. Data were only collected for physical parameters which had been previously identified in the task analysis as directly posing a problem for the operator at his workplace. Noise appeared to be a problem common to all the presses, while lighting conditions appeared inadequate at certain workplaces and thermal conditions near the doors to the outside of the building were also indicated as uncomfortable.

Noise measurements. The global noise levels were recorded at all the presses while the workshop was working and no value under 88 dB(A) was recorded. Another series of recordings was made for each individual press after the workshop had shut for the day. The minimum, maximum and the peak levels were noted for each press at a point usually occupied by the operator. Examples of the noise levels, in dB(A), were:

Press No. 8	88—96—114
Press No. 6	97—105—107
Press No. 10	100—105—110

Certain operators were given a dose meter to wear for the eight hour shift:

An operator on No. 2 press
The meter recorded 360%, which is equivalent to a continuous exposure to 99 dB(A) for 8 hr.
An operator on an automatic press No. 3
The meter recorded 540%, which is equivalent to an exposure of 102 dB(A) during 8 hr.
Maintenance operator
The meter recorded 70%, which is equivalent to a continuous exposure of 87 dB(A) during 8 hr.
Press setter
The meter recorded 150%, equivalent to a continuous exposure of 87 dB(A) during 8 hr.

A number of tape recordings were made which were used to carry out spectral analyses for the noisier presses.

Lighting measures. Illuminance measures were taken at the work surfaces (120–300 lux) and at the press tool (30–50 lux), where visual checks were frequently made.

Thermal environment. Operators had complained of feeling the cold and draughts at the workplaces situated near the exterior doors. The appropriate parameters were measured, for an outside temperature of 7°C. At one position the air temperature was 15°C, the relative humidity 51%, and the air speed 1 m/sec.

Results of the analysis

The results from this study were reported in two stages:

> First a report was written and sent to the firm;
> Following this, the ergonomics research team returned to the firm to discuss the report.

The study report

The introduction consisted of a summary of the project's objectives, which were to provide ergonomics information necessary for the construction of a new press workshop where the working conditions would be better adapted to the working population.

A chapter entitled 'Methodology' was included which presented the approach used by the ergonomists in the study. This was followed by chapters on the task analysis results for the three categories of press (manual, automatic, semi-automatic). A general outline of how to improve the working conditions of the operators on each type of press was given. For example, for the manual press the recommendations covered workplace design and the limits inherent in the machine, reductions in physical work load (lifting, operating press controls), critical incidents and their consequences, allocating longer cycle times to avoid speeding up after incidents and to allow sufficient time for inspection and checking the machine. Similarly, it was suggested that it would be preferable to avoid a 'production line' organization of the presses, as this introduced a number of additional constraints to the tasks due to the interaction of one workplace with another.

A number of recommendations were made concerning the environmental conditions in the workshop. The noise problem was particularly emphasized, both in terms of the risk which certain workplaces presented to the operator's health and the interference with the operator's task due to neighbouring presses masking certain auditory information. Suggestions were made on how noise could be eliminated at source, such as: compressed air leaks; possibilities of installing acoustic insulation which did not interfere with the task, such as enclosing certain parts of the machines; installing shock-absorbing systems to reduce vibration; reducing sound reflections by treating the walls and ceilings; placing the machines further apart and keeping them a greater distance from the walls. The exposure times for the personnel could also be reduced by providing a soundproofed rest room.

The layout for each type of press had been designed to take into consideration the operator's movements, storage (raw materials, tools, etc.), and the movements of the handling equipment (see Figure 13.9).

Workplace layout

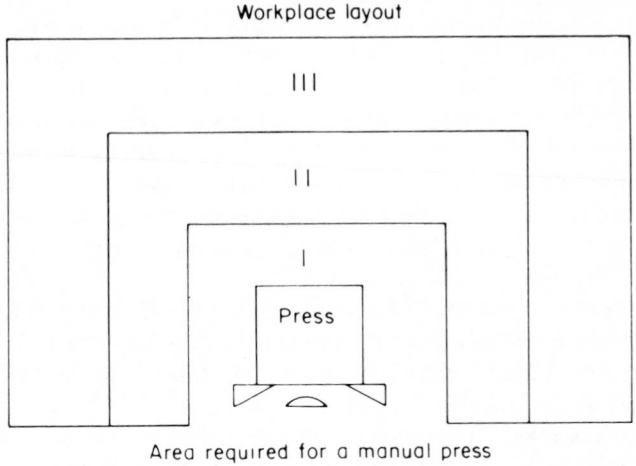

Area required for a manual press

(a)

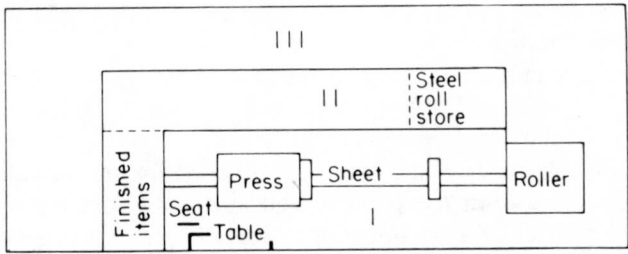

Area required for an automatic press

(b)

Figure 13.9   Suggested total workspace for two types of press. (a) Schematic diagram showing the installation for a manual press. I, area used by operator; II, storage area for raw materials; III, area where fork lift trucks circulate (movement forbidden behind operator). (Areas should be marked out by white lines on the floor.) (b) Automatic press. I, area where operator circulates; II, area where raw materials, etc., are stored; III, area where fork lift truck, etc., circulates. (Areas should be marked out with painted lines on the floor.)

A large part of the report was reserved for the chapter on the conclusions from the study.

First it was shown that there were interrelationships between different problems. For example, moving a press away from a window reduced the reverberation but also decreased the amount of natural light available at the workplace. A number of precise recommendations were also given on how to reduce noise, defining certain architectural aspects of the new workshops, and on the choice of the type of press to be transferred.

A section was also written on the task content and particularly stressed the possibility of 'task enrichment', a point which had been specifically requested in the project proposal. The general task analysis had demonstrated the importance of the operator's control activities at all levels of the task (machine functioning, lubrication, condition of raw materials, quality of the finished product). The proposal for task enrichment was that the checking activities should be recognized as part of the task and that operators should be given the means to assume them efficiently, including being provided with specific training.

The last section concerned recommendations on the safety systems on the press (guards, two-hand controls, photoelectric cell switches). A number of proposals were made which should be taken up with the press manufacturers.

The area occupied by a manual press can be calculated by adding the overall press dimensions and the dimensions of the different zones:

$L =$ Press length + 5 m.
$I =$ Press width + 7 m.
$S =$ (Press length + 5 m) × (Press width + 7 m).

It is possible to calculate the area required for the installation of an automatic press from the press dimensions, around which a 1 m strip should be provided for the operator to move in plus a further 1 m strip for storage. Another 1.5 m width area around the press should be provided for the access of fork lift trucks and other material handling equipment (if movement in only one direction is necessary).

The surface required can be calculated as follows:

$L =$ Press length up to reel mechanism + 4.5 m
$I =$ Press width + 4.5 m
$S =$ $L × I.$

Comments on the report from different sections of the firm

A number of copies of the report were sent to the firm's management. The management passed on copies to certain department heads (Factory

Manager, Methods Department, Planning Department, and Safety Officer), to the union stewards and the shop-floor representatives.

As had been previously arranged, the management invited the ergonomics team to visit the firm to comment on and explain certain points in the report and a series of meetings was arranged.

The first meeting was held with the firm's directors and the department heads. During the meeting a number of criticisms were levelled at the ergonomics analysis of the work organization, notably the measures concerning the operational cycle. A confusion had arisen in the terminology used to describe the work rate of the operators and the time the machines were running. The management had termed the report 'dangerous' over these points. The ergonomics team replied by showing that the data had been collected objectively and that analyses of this type were necessary in order to be able to evaluate the operators' work load. Further, this approach followed from the request to provide information for improving work conditions and introducing task enrichment.

It had to be explained that the proposals concerning cycle times were aimed at providing better production checks on the product and the machines. These tasks were already being carried out by the press operators but in a very restricted manner. It was assumed that the project's recommendations would lead to an improvement in productivity in terms of product quality, decrease in the number of rejects, and better use of press tools. The report was discussed point by point during the meeting.

A programme for implementing the project's recommendations was established which fitted in with the overall programme for building the new factory:

— Building design;
— Machine installation;
— Internal layout of the factory building (offices, washrooms, restrooms, warehousing, corridors, . . .);
— Workplace design for the new workshop, also some improvements for workplaces in the old workshops.

A number of specific proposals were decided on for each of the above headings.

Another meeting was called, with the management, the union representatives and the shop-floor representatives participating. The management explained their plan to improve working conditions and their interest in the contribution ergonomics recommendations could provide.

The union representatives made it known that they would use the report to put forward their claims for improvements to working conditions and to present a case for improving the skill categories of the operators on the basis of the results from the task analysis. The shop-floor representatives said that the report accurately described their work on the presses—one operator remarked that he recognized in the various descriptions in the report the work he carried out and the environmental conditions in the workshop. They also suggested that recommendations for improving working conditions in the new factory should also be carried out in the old factory.

The third meeting was held with just the union representatives and shop-floor representatives. A number of points in the report were explained to the meeting by the ergonomists. One union representative remarked that the improvements suggested in the report were relatively limited and that essentially the workers would remain workers. He would have appreciated an ergonomics study with wider-ranging effects. At this point it was necessary to emphasize the limits of an ergonomics study and remind those present of the limited influence the ergonomics team had on implementing any improvements in working conditions.

A meeting was also held with the factory manager. The points covering the building design and machine installation were put onto the project plans. Propositions for reducing noise had been studied: acoustic treatment on wall and partitions, distancing machines, separating noisy machines by storage areas, using insulating enclosures, silencers on compressed air outlets, mounting presses on anti-vibration supports.

The ergonomics team considered that it had been useful to hold these meetings as they provided an opportunity of explaining and evaluating the study. The opportunity to comment on the study report should be integrated into the methodology of industrial ergonomics studies.

Evaluating the impact of the ergonomics study

An evaluation study was carried out approximately one year after the new factory had been put into service. The same group of ergonomists was responsible for the evaluation as for the original study. The observations were carried out in two days and were intended to estimate what ergonomics recommendations had been used and to compare the old and new working environments.

The same methods were used to analyse the tasks in the new factory as had been used in the original study described above. A comparison of the two situations, old and new, was then possible using observational data from the two studies.

However, it should be noted that any comparison has limited validity, as the whole situation had evolved and not only those points that had been studied previously. For instance, some changes had been made in the production techniques (semi-automatic presses had been taken out, and the automatic presses had been developed and modified) which made any direct comparisons with the old factory difficult. Similarly, other factors had changed which had not been covered by the ergonomics recommendations. It should also be emphasized that it is very difficult when comparing two work situations to isolate improvements or changes which are attributable solely to ergonomics rather than other factors.

With the hope that some sort of objective comparison could be made only two types of factors were considered:

— The physical environment
— The task content within the job.

The physical environment and material conditions in the new factory

The conditions in the new factory had been considerably modified. The ergonomics recommendations had been taken into account and solutions for most of the problems had been found. The ways in which each problem interrelated with others were also considered.

The main modifications involved the building, the use of space, improvements in the physical environment, and individual workplace layout.

The building. The building architecture had taken advantage of all the possibilities for the use of natural lighting suggested in the report. The south west face used windows with a swivelling chassis and blinds, windows on the north west face were installed on the upper part of the wall, and 10% of the roof was translucent.

The internal wall surfaces had been covered with an acoustic insulation which was also relatively easy to keep clean. The walls also contained a layer of fibre glass which provided good thermal insulation.

The floor was clean and smooth, and corridors in the main workshop had been marked out with painted lines.

Interior layout. The area provided for each press was slightly smaller than that recommended on the basis of the task analysis (provision of raw materials, storage, maintenance, and adjusting the press). The length of the building had been increased by 10 m, which provided 250 m² of additional space.

All the service buildings were built with soundproofing: a restroom for the press operators, the canteen, offices, quality control and testing laboratory, washrooms, and medical centre.

The corridors used by fork lift trucks were separated from the corridors used by personnel, which were situated around the outer portions of the workshop.

The proposal that each press should have its own storage area had not been accepted. The management had chosen to provide a central storage area serving all the presses. However, the overall installation provided easy access for the fork lift trucks to each press. It was still difficult to organize raw material storage around each individual press as the presses were very close to each other, but the compromise chosen was acceptable.

Noise problems had been reduced by introducing sound insulation techniques which will be discussed below, and the use of the restroom allowed the workers to communicate freely.

Physical environment

Noise. It is now possible to talk in the workshop at normal voice level without having to shout. The measures taken to reduce noise had resulted in a considerable decrease in noise levels (acoustic coverings on partitions, silencers on compressed air outlets, enclosures fitted to the noisier machines).

The majority of recommendations from the report had been taken up, although there still remained room for improvement on certain points. The firm had invested considerable effort in this area and should be complimented on the thoroughness with which they approached the problem of noise in the workshops. A group of acoustic consultants employed by the firm used the noise measurements from the old workshops to predict likely levels in the new building. On the basis of these calculations a noise insulation programme was planned to reduce noise to acceptable levels. One of the decisions taken was to enclose certain very noisy machines and manufacturers of enclosures were contacted. They were given specifications covering the level of attenuation required for the workshop and these also included a section on design modifications necessary for production and maintenance requirements (input, output, adjusting the press, cleaning, visual requirements, access to the press).

The small presses were installed on shock-absorbent blocks and the larger presses on vibration-absorbing foundations.

The firm purchased the necessary equipment to carry out noise measurements. The industrial safety department set up a file of acoustic measurements around each press. The measurements were made on different occasions, with the presses working under different conditions, using different tools or making different products.

The results from the acoustic measurements during the evaluation showed that the changes adopted were effective in reducing noise levels in the new building:

Figure 13.10 Photograph of the new press workshop

Restroom   60 dB(A) minimum, 66 dB(A) maximum
Offices   57 dB(A) minimum, 61 dB(A) maximum
Enclosed presses   69 dB(A), that is to say a 26 dB attenuation over the
                    same press in the old workshop

Lighting. Natural lighting provided illumination levels of over 700 lux for most of the time. Artificial lighting from banks of fluorescent tubes had been installed to give a uniform illumination of 300 lux. The fluorescent lighting was switched on automatically by a photoelectric cell which ensured that illumination did not go below 300 lux.

Individual lights had been installed inside the acoustic enclosures on the automatic presses. However, this had not been entirely satisfactory as there was a considerable difference in levels of illumination inside and outside the enclosures and reflections in the enclosure windows tended to be annoying.

Taken generally, the increase in illumination had been an improvement for the press workers and drivers of the fork lift trucks, as well as for the cleaning and maintenance operators.

The people working in the new building remarked that they appreciated being able to look outside as a result of the windows being installed at eye level.

Thermal environment

The building had the benefit of efficient thermal insulation. A double-doored exit had been provided to eliminate draughts in the workshop. The heating system maintained a comfortable environment in winter, but in the summer the workshop could become too hot due to radiant heat from the window surfaces in the roof and heat from the presses.

A method of reducing the workshop temperature by pumping cold water through the heating system radiators was under consideration. An experimental period showed that the ambient temperature could be reduced by 3–4 deg C.

The improvements in the thermal environment had contributed to the workers' greater comfort and also to better machine use; the number of tools broken and the time taken to adjust the hydraulic systems had both been reduced.

Modifications to workplace layout

Seating. The stools had been modified and the operators had seats which could be moved and adjusted. On the manual presses it was the design of the press which prevented the operator taking up a correct posture, because there was no room for his legs under the press front.

Work surface. Custom-made work surfaces, adjustable in height, were designed by the firm to equip all the different work spaces.

Technical improvements. A number of improvements were implemented to eliminate certain serious constraints on the operators' work. A conveyor belt was introduced to take away metal waste from the press, smaller containers were obtained which were lighter to handle and allowed faulty items to be sorted out more quickly, and the lubrication system was modified so that it did not require constant surveillance and no longer shot out jets of oil.

Safety systems. Guards made of a transparent material had been installed on the machines. Now the safety equipment is systematically checked by the operators each time before they start work at the beginning of the shift.

The safety system had been modified; on certain presses it had been taken off altogether after the introduction of a mechanical arm for automatically transferring items to the press. This equipment presented certain advantages such as increased safety and better postures for the operator, but there were also certain disadvantages such as increasing the number of different items which had to be checked and the possibility of introducing further types of critical incident.

Task content

A short task analysis was carried out at each of the different types of press so that the old and the new workshops could be compared. The subjective impression when visiting the new workshop was that the task itself had not changed but was being carried out in better conditions.

The analysis showed that automation in the factory had been significantly developed, the number of automatic presses had increased, replacing manual presses, and new types of automatic control systems had been installed (feeding the sheet metal, lubrication, clearing away waste, . . .).

One manual press had been equipped with a mechanical transfer arm and two others with an automatic handling system for feeding the machine and positioning the items.

The press operators' tasks on the different types of presses had changed, and the results from the analysis have been divided into automated and non-automated presses.

Task content at automated workplaces. The manual presses equipped with a mechanical arm or with automatic handling had improved the safety and work posture of the operator but the surveillance task was more demanding: certain items were not picked up for positioning and quality control checks on finished items were carried out periodically in the containers rather than, as previously, item by item as they came off the press. There was insufficient time available during the evaluation to check the type and frequency of critical incidents in pick-up, positioning and ejection operations or to observe the operator's point of regard during these processes.

The continual stopping and running of the automatic presses had been reduced as the technical problems had been solved. It could be expected that the operators were less occupied, but the systematic observations showed that visual checks to the steel-feeding system and the lubrication system were less frequent but that the checks towards the item as it left the press represented 67 % of the running time, against 73 % in the earlier workshop. The difference can essentially be considered as qualitative. In the old factory visual checking was aimed at anticipating any of the frequent incidents which arose, but in the new factory the control task consisted of detecting rare incidents.

The production levels on the automatic presses had increased as a result of the improvements made to the feed mechanisms and running of the machines. But as the items came off the press there were no guides or storage mechanisms. The operators were faced with increased workload for organizing and checking the items. Consequently they were then obliged to stop the press for longer periods than those previously necessary.

Task content of non-automated presses. The work on the manual press had not been changed but certain modifications to workplaces had lightened the demands of the task. The lighting had been improved, the transfer of items had been made easier by installing a small shelf, and the safety system recommendations had been introduced.

Three workplaces were still arranged in a production line but this was only a temporary measure. The organization of the production line had conformed to certain recommendations made in the report. For instance, the first press had been replaced by an automatic press which was controlled by the neighbouring manual press in line. There was an increase in throughput with this arrangement, but the operator working on the manual press had the extra load of checking the operation of the automatic machine.

The press operators appreciated the introduction of the automatic press but were not happy about the consequences to their own tasks, where cycle times had decreased and the interdependence between workplaces remained unchanged.

A number of points should be considered in relation to the evaluation phase of the ergonomics study:

—   It is not easy to compare two situations using ergonomics criteria, as the different changes have different effects on work load (transferring mental and physical activities).

—   Production systems are dynamic and continually evolving, whereas

the original study and the evaluation were carried out at fixed points in time.
— New problems may well be brought to light during the evaluation stage and new sets of recommendations given.
— Other variables (socio-economic, financial, etc.) also influence the working environment and should not be ignored during the evaluation.

Conclusions

There were two main reasons for carrying out this study:

To demonstrate the usefulness of applying ergonomics in industry;
To develop the methodology of applying ergonomics in industrial projects.

This chapter has attempted to present the methodology used in the study as clearly as possible. Considerable importance has been given to the evaluation phase, which in the study team's experience is the phase the least often completed in a project programme.

In conclusion, the following points should be noted:

—The majority of the recommendations concerning the physical working conditions and the physical environment were taken into consideration in the new building. The actual solutions accepted were, of course, the result of a compromise between ergonomics criteria and other requirements (technical evolution towards automation, economic costs, . . .). This should not detract from the considerable improvements made to working conditions in the new factory.

—The recommendations concerning the operators' task content, work organization, and training had not, at the time of the evaluation, been put into effect. However, a project on training for quality control and reassessing operators' skill categories had been developed and was soon to be started. The management considered that a first step in the right direction had been made and this demonstrated their willingness to carry through the other recommendations.

—The problems which the ergonomics study had brought to light stimulated lively debate at all levels of the firm's hierarchy. Ergonomists have to accept that their methods and the results provided by them provoke certain contradictory arguments. It is not possible to simply impose scientific reasoning on a socio-technical system which contains a number of divergent interests. In particular, a number of questions had been raised concerning the reduction of manpower with the introduction of automation and the resulting increases in work load. It is not necessarily within the ergonomist's realm of competence to answer such questions but he cannot ignore that changing technology can influence levels of employment.

If the approach adopted had been carried through to the full the research team should have assisted while the recommendations made in the first part of the study were put into practice. However, this was not the case as the team did not return to the factory between the time the first report was presented and the visits for the evaluation study in the new factory.

A positive aspect of the ergonomics study was the willingness on the part of the management to derive solutions to particular problems, sometimes with outside help as with the noise problems. However, the study team felt that they should have been consulted on certain points of the layout modifications (seat adjustment, press control positions, and placing the individual lighting).

The evaluation demonstrated both the positive aspects of the ergonomics study and its limitations. The management, personnel and unions all recognized that the study had contributed to improving the working conditions.

Finally ergonomics, with its approach centred around the study of man at work, provides a means of adapting technology (buildings, machines, tools, products), working methods (organization, training), and environment (noise, lighting, thermal conditions) to working man.

Note

The study presented in this article was carried out by a group of research workers from the Laboratoire de Physiologie du Travail et d'Ergonomie du Conservatoire National des Arts et Metiers. The laboratory was commissioned to do the study by l'Agence Nationale pour l'Amelioration des Conditions de Travail.

One of the functions of this agency is to promote pilot projects which are aimed at significantly improving working conditions.

The Laboratoire de Physiologie du Travail et d'Ergonomie contracted to do the project with two objectives: (a) to demonstrate that an ergonomics study could contribute to the improvement of working conditions; (b) to develop research in the methodology of ergonomics studies in industry.

M. Jankovsky is now Chargé de Mission at the Agence Nationale pour l'Amelioration des Conditions de Travail.

The figures in this article are reproduced by permission of ANACT. The article itself is based on three reports by J. Duraffourg, F. Guerin, F. Jankovsky, and J. C. Nascot.

1. *Analyse ergonomique du travail dans un atelier de presses en vue du transfert de certaines presses dans un nouvel atelier à construire.* ANACT, Paris, 1976, 92 pages.
2. *Analyse des activités de l'homme en situation de travail. Principes de méthodologie ergonomique.* CNAM, Paris, 1977, 122 pages.
3. *Une intervention ergonomique.* ANACT, Paris, 1979, 97 pages.

as a production factor had evolved into a 'critical mass' over a long period of time. It only needed a trigger.

The 'towards codetermination in work life' period at the end of the 1960s and during the 1970s

A major field of research evolved at the end of the 1960s, viz. the democratization of working life. A comprehensive programme of research and trials was conducted in many sectors of work life: private companies, state-run companies, national, municipal and county council authorities, consumer cooperation, etc. The authorities in charge of research and experiments were delegations and committees representing the different parties in the various fields of working life. These groups contacted the scientists.

Other research fields which arose as a result of the new subject categories were as follows:

1. Behavioural toxicology—the study of the effects of solvents and other chemical substances (found in the working environment) on man's mental functions. Here it was found, for example, that various mental disturbances can serve as warnings of hazards in the work environment at an earlier stage than physiological changes in man's body. Warnings studied involved reduced alterness and impaired reaction ability with a greater risk of making errors and/or residual psychophysiological damage as a consequence.

2. Psychophysiology—the study of the effects on man, using psychological tests of function, of different loads in work, such as physically exhausting work, work in heat or cold, vibration, noise, radiation, and other environmental factors capable of exerting an influence on the central nervous system and of which the effects can be revealed by means of psychological and psychophysiological methods.

3. Behavioural science research into accidents is currently critical of the theory of 'accident-prone' people. Instead, attention is being devoted to the interplay between work environment, work duties and human reactions. Accidents and near-accidents are being analysed in terms of disturbances to or deviations from planned procedures.

4. Research into the work duties and design of work organizations has been intensified. One example of fresh research discussion in this field is that of whether there are sufficient research results to warrant condemnation of certain organizational forms, viz.

— Extremely specialized, repetitive work in short cycles;
— Control of the work process through technology which prevents people from being able to affect their work pace or way of doing the job;

— Work design which makes it impossible to or prevents contact with other people during work;
— Certain forms for work alone;
— Utilization of incentive pay for dangerous jobs.

5. Research into work supervision has concentrated more on the effects of the total management system (decision-making process, degree of delegation, group organization) than on individual authoritarian or democratic leadership abilities respectively.

And in another research field (6), attempts are being made to elucidate the negative effects of passive adaptation to work life, viz. the adaptation which takes place when the individual screens himself off from poor working conditions in order to reduce their effects on ego perception. He denigrates the importance of work to his situation in life and to his mental health.

A connection has also been found between passivity as a consequence of pacifying activities, political passivity in civic contexts, and passivity in leisure time.

Research becomes controversial

There is a tendency for scientists to identify themselves with their research results, and they often feel that it is their duty to engage in active dissemination of these, which often reveal dangers to physical and mental health in many of our present work environments.

Occupational research has thus become more controversial. A recent study among scientists disclosed that 56 per cent of the respondents had suffered reprisals from decision-makers on the labour market. Twenty-four per cent had been 'stopped at the door' and denied admittance to work sites they had to visit for research purposes. Seven per cent had been barred from work sites during ongoing work site investigations. Twenty-two per cent had been denied requested information about the work site. The reports of 12 per cent had been censored and 11 per cent of the reports had been classified as secret.

However, events such as these can be regarded as passing symptoms during a period of transition in a country in which reform efforts during the past few electoral periods have entailed debate on industrial and economic democracy and in which new legislation in the labour market field has made possible research in areas long in need of it but lacking the formal qualifications for research grants.

Organization of research and training today

The organization of research is very complex in all research fields, including behavioural science occupational research. Developments in Sweden are currently characterized by rapid growth in sectorial research, i.e. applied research and development linked to different sectors of society. Basic research has by no means displayed the same growth. The growth in applied research has led to a need for better contacts between the research and the consumers of research results. These consumers were previously found mainly in the private industrial sector and in state and municipal authorities. But contacts in recent years have been expanded to include trade union representatives, which has also contributed to the establishment of new research centres. In 1970 the Occupational Medicine Unit of the National Board of Safety and Health was supplemented with an Occupational Psychology Unit. The Swedish Work Environment Fund was established in 1972 and is currently the most important source of grants for occupational research. In 1977 a new research institute, the Swedish Center for Working Life, was opened. It has currently about thirty scientists.

Even closer contacts with trade union organizations have been proposed, and the unions themselves have demanded greater codetermination in the choice of research fields and research trends.

The research is financed in different ways. A scientist may receive a grant:

1. From Governmental Research Councils governed by selected scientists.
2. From ministries, which are government-controlled bodies. Ministries finance research either (a) through grants (e.g. *via* the Delegation on Social Research or to the National Institute for Building Research, which allocates further) or (b) from government agencies and other institutions (such as the Swedish Center for Working Life).
3. From the Bank of Sweden Tercentenary Fund (which was a main contributor of grants to behavioural scientific research in the 1960s). The Bank of Sweden Fund is governed by 16 members of parliament and 16 scientists appointed by the parliament.
4. From independent funds, such as the Swedish Work Environment Fund which has a board where representatives from the labour market are in a majority, although the employers hold more seats than the employee representatives. The Fund is financed by taxes paid by the employers in relation to wages and salaries paid out.
5. From a large number of small funds of varying origin, often private donations.

There are a number of interesting details in this development. The employers' behavioural research within the occupational field measures date from the 1950s, when they founded the Swedish Council for Personnel Administration, the main centre for this research in Sweden in the 1950s and 1960s.

However, polarization in behavioural scientific occupational research resulted in the assumption by political and trade union forces of control over research through the starting of new funds and institutions. During this time the Swedish Employers' Confederation expanded its own Technical Department. One way to put it is that influential directors in the private business sector realized the importance of behavioural scientific research at a very early stage, especially research conducted in conjunction with the human relations movement. To some extent, however, the private business community in other western countries actively opposed every attempt to implement the results of this kind of research. So employers in Sweden can be regarded as pioneers in this field.

As mentioned above, however, there was a strong polarization in 1967/68, leading to a shift in power. An example of this polarization was the Mental Health Campaign conducted at the end of the 1960s. This was financed by the employers. The campaign outraged political and trade union opinion, which savagely criticized its criteria for mental health. The criteria were described as an attempt to tailor people to the prevailing system on the labour market. The fact that the design of work tasks, the structure of work organization, the structure of influence and technical and administrative systems were all regarded as fixed for all time and that behavioural science would be utilized to tailor people to these structures so that adaptation would take place with the least possible effect on the mental health of staff and employees was sharply criticized.

Training of scientists and working environment staff

The education of scientists takes place *via* the usual channels of Ph.D. training at the universities. Doctoral students apply to research groups attached to their departments which are conducting studies of interest to the prospective scientist.

Whether a person has become a scientist in the field of occupational research has hitherto been more of a coincidence in the choice of a university department at which he could begin his basic training, the research groups to which he had access there, and the influences he experienced during that basic training. A new phase in the training of scientists was initiated in conjunction with the opening of the Swedish Center for Working Life. Here attempts are made to interest white-collar and blue-collar workers in becoming scientists. Contacts are made *via* the trade union organizations and aim primarily at people actively involved in union activities. The basic idea is that the wide experience these people have of working problems and their excellent insight into the realities of working life give them a big lead over scientists who have only followed an academic career. To begin with, these employee-scientists would serve as project members in research projects. It is hoped that at least some of them would subsequently acquire formal scientific training.

If the recruitment of scientists is to some extent left to chance, the training of white-collar and blue-collar workers who are employed or would like to be employed in departments of companies and agencies conducting work environment programmes is being carried out in a much more efficient manner. After a basic training in psychology in Stockholm, a student can select a postgraduate line comprising occupational psychology up to one-eighth of the total training. It is also possible to take detailed courses in special subjects.

Union safety supervisors are trained *via* their union organizations but can also apply for supplementary training at the departments mentioned below. Safety engineers can obtain supplementary training *via* their professional organization by means of courses run by the National Board of Occupational Safety and Health and in different programmes under the auspices of the Swedish Employers' Confederation. Company doctors and other staff of company health centres receive supplementary training from the National Board of Occupational Safety and Health. Other organizations providing courses are the state-run training centre (Statskonsult), the Swedish Council for Personnel Administration, which holds a few courses in working environment, and the Institute for Applied Psychology in Saltsjöbaden. However, most of the supplementary training is provided by the National Board of Occupational Safety and Health and the Swedish Employers' Confederation's Institue for the Training of Foremen.

Main centres for behavioural scientific occupational research

The main centres outside the universities are the Work Psychology Unit of the National Board of Occupational Safety and Health, the Swedish Centre for Working Life, the Laboratory for Clinical Stress Research, the Institute for Applied Psychology at the University of Stockholm, the Institute of Applied Psychology at the University of Gothenburg, the Working Environment Laboratory of the Royal Institute of Technology, the Research Institute of the Swedish National Defence, the Institute of Technology in Luleå, the Departments of Psychology and Sociology at the University of Gothenburg, the Department of Psychology at the University of Stockholm, and the Government Clinic for Occupational Tests. In addition to these institutions, the Swedish Employers' Confederation has its own Technical Department, although the department does not conduct much research but engages in employer-oriented pilot programmes.

A description follows of some of the most important centres, which, more than others, have shown interest in occupational research:

1. *Work Psychology Unit of the National Board of Occupational Safety and Health*
Has about fifteen scientists. The unit's function is to conduct, monitor and stimulate research into the social, mental and psychophysiological consequences of the total working environment. This environment may involve physical, technical, organizational, psychological or social aspects. The unit devises methods to improve working environment efforts in companies and authorities.

2. *Swedish Center for Working Life*
Has about thirty scientists. Its task is to support reform efforts in working life by means of research, development, and information. Its research aims at promoting the development of a good working environment featuring equality of influence, opportunities for all-round development of employee interests and skills, respect, security, and good material conditions. The Occupational Research Centre is a central reference agency for applied research and development work in the field of working life and is responsible for a reference library which provides complete coverage of research and development work in occupational fields in at least Scandinavia.

3. *Laboratory for Clinical Stress Research*
The Laboratory is attached to the departments of medicine and psychiatry of the Royal Caroline Hospital in Stockholm and has about thirty scientists. Studies of individual functioning and well-being during different types of loading, such as stress responses to under- and overloading in working life with respect to both mental and physical loading, are found in the research programmes.

4. *Department of Psychology, University of Stockholm*
The main emphasis in research in the section concerned with the social psychology of working life involves the study of relations between different aspects of work and the work environment and the individual's perception and health.
   Another section is concerned with psychobiological research, mainly with effects from under- and overstimulation.

5. *Department of Psychology, University of Gothenburg*
Research has been conducted into routines for personnel administration and comprised a study of executive duties in the civil governmental apparatus with a view to improving the training of managers. Other projects have involved the design of staff policy programmes in companies and employee attitudes to such programmes. Research in codetermination is also on the programme.

Examples of current research fields

For the sake of simplicity, I shall refer to the numbering of the main centres as used in the preceding paragraphs when I wish to indicate where research in one of the fields mentioned is being conducted. It should also be noted that an interdisciplinary approach is becoming increasingly common in occupational research.

Physical and chemical loading factors

At present, considerable research is being conducted in these fields in order to ascertain the effects on the human organism of physical and chemical loading factors in the working environment. Examples of such factors are cold, heat, vibration, electromagnetic fields, heavy lifts, etc.

Acute and chronic effects on the nervous system are being studied (1) of exposure to neurotoxic substances such as xylene, ethyl benzene, jet fuel, solvents for lacquers and paints, anaesthetic gas, styrene in the plastic industry, etc. For example, behavioural disturbances have been found to arise before a medical examination is capable of disclosing any damage. This means that mental tests of function can be used as alarm systems for unhealthy work environments.

Stress research in under- and overstimulation

Under this heading can be found research which sheds light on the prevalence of psychosomatic disorders as a result of understimulation (e.g. assembly-line work) or overstimulation (e.g. stress due to too wide a variety of duties (3, 4)). The work situation in relation to biological rhythms is also being studied, for example individual circadian rhythms and situations such as shift work (4), displacement of circadian rhythm in airline staff in transatlantic traffic. Another example is a project dealing with the relationship between risks of working alone and the psychological effects of working alone (1).

Accident research

This research is currently being conducted on the basis of elucidation of the interplay between working environment, the design of work tasks, the decision-making structure, habits, and human reactions (1).

Research into work organization

Here, the consequences of different types of work organizations and work designs and their psychosocial effects on individuals and groups of people

(1, 4) are being studied. Other projects are studying the relationship between the degree of mechanization and work satisfaction and mental health respectively (4). The psychosocial effects of wage forms are also being studied (1, 2), as well as relationship of work environment to absenteeism (2, 4).

Research into mechanisms of change

Here the procedure involves studying and drawing up methods for dealing with changes in the working environment (1) and improvements in trade union work forms (1, 2) and the working environment (1) on the basis of educational measures. The conditions and opportunities for conducting efficient working environment measures in companies and authorities are being studied (1).

Research in codetermination

Some projects in this field deal with assessment of the new codetermination legislation (1, 2, 5). Other projects are examining obstacles to and opportunities for implementation of this new legislation (1, 2, 5). A few projects are examining the influence structure in different work organizations (2) and the structure of union competence and organization (2).

Equality between men and women and relationship between working life and family life

Studies are being conducted into the manner in which work perception is affected by different conditions in the surrounding community and labour market (e.g. the existence of child care facilities, housing conditions, double work by women on the job and at home, etc.) (1, 4). Research from this viewpoint is also being conducted into absenteeism and job attendance (4). Other programmes are examining in more general terms the obstacles to and conditions for sexual equality in working life (1, 2).

Perspectives

The new research needs are the product of three trends:

1.   The rapid postwar rise in production. In many cases, the attendant increase in material prosperity and social welfare has been obtained at the cost of drastic changes in working life and the working environment.

2. The importance of the working environment to well-being has aroused increasing interest.
3. Growing demands have been raised for increased influence for workers.

The development of future research in the study of psychosocial issues in working life depends to a large degree on the interest and knowledge of the groups in society with influence over research and research grants. As mentioned above, trade union organizations are tending to demand increasing influence over the choice of research areas. Their wish for greater knowledge is also likely to meet with increasing sympathy. If the idea of stimulating experienced trade unionists to participate in projects and research became attractive (and if any of these people decided to obtain scientific training), this tendency would definitely be reinforced. More issues and problems faced by the working man in his working environment would probably be regarded as 'worthy of research'. It is undeniably true that scientists who have never worked on the factory floor probably hatch research ideas different from those conceived by someone who has spent 15 years on an assembly line.

This perspective has an equivalent in education. For the past ten years in Sweden the *studentexamen* (higher school-leaving certificate) has been abolished as a requirement for admission to university. Today you only need to be 25 years of age and have four years of gainful employment behind you in order to participate in certain university training. This reform has attracted adults to begin school again, a good number with excellent results. So there is no official obstacle for people with vocational experience and practical knowledge of working environment problems to switch to scientific studies in the field of behavioural science occupational research.

The former chairman of the Swedish Metal Workers' Union, now chairman of the Industrial Safety Fund, made the following remarks in an interview: 'Working environment efforts in the 1970s have been largely concentrated on correcting the working environment destroyed in the 1950s and 1960s. This is especially the case for chemical health hazards and asbestos. Research into chemical health hazards has hitherto consumed a large part of the Industrial Safety Fund's resources whereas ergonomics have been allocated a relatively small share. Noise research has cost 20 million kronor. Accident research has just been started and will increase. Ergonomics will be considered at many new work sites, for example at places with computer terminals. One new type of project takes up psychosocial problems'.

In a review of research needs a report from the Industrial Safety Fund noted that there is a lack of research on the higher levels in the company and the company's relation to the community. It was also noted that problems are often selected and tackled on the basis of the company's or authority's (i.e. their leaders') point of view. To date, trade unions have framed research problems or been able to influence the starting point for research only to a limited extent. The issues likely to be studied in the future with behavioural science and other scientific measures are work organization and production systems, supervision and coordination in companies and authorities, working life in relation to society's political, economic and legal structure, and working life in relation to man, the family and the community's social functions.

A great deal of this research will probably be conducted under general headings such as codetermination, equality between men and women, the effects of working life on people's social behaviour, etc.

It can be said that behavioural science research has already shown that working conditions in many professions today have a direct influence on intensive and persistent feelings of monotony, social isolation, powerlessness, being rushed, and fatigue. These feelings are becoming increasingly common in large groups of white-collar and blue-collar workers as work content is objectively impoverished and controlled. But the objective is greater than merely removing adverse elements in working life. Research must point the way to a working life in which people have a true measure of codetermination, experience satisfaction in their work, and are able to attain better life values in more than just the material sense. In closing, here is a brief outline of research areas drawn up by the Industrial Safety Fund after questioning scientists and labour interests about the need for, interest in and plans for research (after Lennerlöf, 1979).

Physical and chemical load factors

The effects of solvents on central nervous system functions will be an important research field for years to come. The aggregate knowledge in the field to date and method development will lead to close collaboration between company health units in making biological checks on work environments where exposure to solvents is found. An increase in research measures in ergonomics can also be foreseen and conflicts between man–machine systems will be examined to a greater extent.

Stress research into over- and understimulation

Here, projects are being discussed such as (a) the relationship between different types of information and prevalence of psychosomatic complaints, (b) the working environment at university with respect to stress factors, (c) mental strain in retail trade work with a view to a transition to larger units, (d) the reasons for high staff turnover in public health and especially the significance of the working environment, and (e) reasons for increased absenteeism due to illness.

Accident research

In the future, greater emphasis will be placed on studies of the decision-making structure and learning mechanisms. The model for analysis of individual accidents and models for systematic feedback of information to industrial safety workers will be developed. Research will be conducted into man's subjective perception of risk, risk awareness and risk-taking in relation to, for example, sexual role behaviour and risk learning. Analyses of the total work environment are being planned in order to shed light on processes which lead to accidents.

Research into work organization

Research will be intensified into the risks and disadvantages of fully automatic processes, leading to increased control of the working man, and the effect of lack of social contacts because of the relatively small number of people in automated plants, etc. The mental and social consequences of working alone will be studied. The social and mental consequences of technical changes in industry will also be the subject of research. One example here is computerization and the anticipated introduction of assembly-line work in office environments. Steps will be taken to shed further light on different forms of work organization and their effects on changes in vocational roles plus the psychosocial climate in companies.

Research will follow the trend to more group-oriented organizational forms for increased codetermination. Control systems in production will be examined, such as different payment systems as a form of work control. Research will also contribute to the development of alternative forms for work organization and work design towards increased codetermination and self-determination. A very important research field will be the development of methods for measuring the psychosocial consequences of the work environment.

The manager's role as a component of the work organization will also be the subject of study in the future. Management with codetermination will impose new and difficult demands on the management role. Future changes in working life will have a significant effect on the manager's role.

Relations and conflicts in work groups

Research here will examine the psychosocial effects of people's behaviour towards one another in working life. People constitute the social environment for one another. Disturbances in the psychosocial work environment as a result of conflicts in work groups or with management will be studied in a number of projects on the basis of theory derived from family therapy work. Conflicts and conditions for conflict resolution and mechanisms for rejection of individual group members will be investigated. The forms for mental punishment and social control will be studied.

Research into the process of change

A number of experiments will be conducted in collaboration with interested companies and local unions using new forms for work environment activities. Other efforts will be made to find and test various models for information dissemination and activation in work environment and industrial safety work. There will also be a systematic reinforcement of knowledge with a view to increasing employees' opportunities to influence the design of their work environment. More research will be devoted to the processes which lead to increased codetermination. More general research into change will also be conducted. New methods and instruments the employees can themselves use in accomplishing change will also be developed.

Codetermination research

One of the tasks of research will be to shed light on the various obstacles to employee codetermination. The conditions for democratic decision-making processes in working life will be examined. The need for changed attitudes in individuals in attaining democratic processes will also be studied.

Development of codetermination for employees will also impose new demands on trade union organizations, even leading to a need for new work forms in union activities.

One new function, the employee consultant, is being developed and will be studied in the future. Another research field will be the study of obstacles to and opportunities for immigrants to participate in industrial democratic processes.

Equality and relationships between working life and family life

Major research efforts will be made for the purpose of improving the quality of life for people. Studies have disclosed close links between pacifying work life and passivity in political participation and leisure-time activities. For women, family work has entailed sole responsibility for care of the home. Issues related to obstacles to and opportunities for achieving equality between the sexes will be studied from this perspective. The need by men for greater family orientation will lead to research into the career thinking caused by sexual roles in working life.

Working hours will also be studied from different points of view, for example with regard to the relationship between the location of working hours in the day and the family–social effects.

Need for codetermination in the choice of research fields

Major changes are expected regarding the way research is initiated and the problems research will examine in working life. When the first power-saw was introduced into forestry work decades ago, a reporter did a story on the event and asked one worker: 'How do you feel now? Isn't it wonderful getting rid of all that back-breaking work?' 'Sure', answered the worker, 'but I freeze like hell.'

No reporter, no company director, no researcher grasped what that worker meant. For them the answer seemed silly. *They* did not understand the problem. *They* had not spent time in the forest cutting down trees. *They* did not understand. And the worker lacked the concepts to explain the nature of his problem.

We now know that vibration causes physiological disturbances in function, in this case in the hands of the worker, which felt cold. Occupational research must achieve (and is certain to do so) closer contact with the naked realities experienced by the working man in his work environment.

Reference

Lennerlöf, L. (1979). *Arbetsmiljön ur psykiskt och socialt perspektiv.* Föredragsmanus, Stockholm.

Stress, Work Design, and Productivity
Edited by E. N. Corlett and J. Richardson
© 1981 John Wiley & Sons Ltd

Chapter 15

Answering Workers' Needs in the Design of a New Factory

J. M. Leduc
Federation Generale de la Metallurgie,
Confederation Française Democratique du Travail
Paris, France

The FGM–CFDT welcomes the opportunity to set out its views on the way in which new factories should be designed in order to better satisfy the needs of workers; this will be a representative point of view from the union.

We have agreed to give our organization's position because: first, it is, in fact, the workers who are essentially affected by the design of factories in which they must spend a major portion of their lives, and we know we have to find our own solutions to the main problems; second, there has been a long history of cooperation between the Laboratoire de Physiologie du Travail CNAM and union groups at various levels of our Federation, and we are pleased to say that this cooperation has proved very fruitful for us.

However, the subject we are about to discuss raises certain fundamental questions which we must consider at the outset: employees do not choose their work; working conditions cannot be separated from living conditions; a piece of equipment cannot be judged independently of its utilization; the organization of work always means the division of labour; new factories, but to what end, and to whose advantage?

Such questions may seem rather philosophical. In fact, we have to consider points like this because they largely explain the difficulties that Philips or Renault encounter when installing a factory in India; they are not only selling the means of production for cars or electronic components, they are selling a way of life—a capitalist type of development. The questions posed above will not be resolved by just changing the orientation of the windows or the equipment layout.

After clarifying each of these points, we will indicate the direction our inquiry might take.

Essential observations

Employees do not choose their work

The capitalist company chooses its personnel, selecting it according to criteria which will assure immediate profitability. The choice is particularly easy since the job market is organized by these same capitalists; there is always a surplus demand. In the present employment crisis in France this situation is very apparent: 1,200,000 unemployed, not counting all those who do not take the trouble to register on the unemployment rolls and the 500,000 partially unemployed. There are only too many to choose from, but even outside this period of acute crisis the phenomenon is the same: capitalism has the resources to impose its own working conditions.

The transfer of technology to developing countries is not done for humanitarian reasons. Companies locate where they can find the cheapest labour: in Puerto Rico, Mexico and Hawaii for American firms, who otherwise have to compete with Europeans and Japanese in the Far East and Africa, in Lower Normandy and Brittany. Factories are installed in the areas where there are no other jobs available and workers become prisoners of their work stations, whatever the working conditions may be.

Companies may utilize a floating labour force: the immigrants who come from every country, with their own customs and language, and who are even less able than others to give their opinions on their future working conditions—for instance, the North Africans and Portuguese at Chausson, at Cables de Lyon and at Ideal-Standard (today, most of them are unemployed), the Turks at Peugeot, the Yugoslavs in the Lorraine steel industry, etc. These examples are restricted to those firms where recent industrial unrest has been sorted out.

However, locating plants abroad can pose certain problems (political uncertainty, transfer of personnel, transportation and communication costs, repairs, etc.), while relying on too large a number of foreign workers leads to socio-political situations which are difficult to control. Furthermore, the French Government, following other European governments, is heading towards a policy of restricted immigration. The solutions for replacing these workers are available.

One solution is to rely even more on the reserve of unemployed workers. The French 6th Plan had set the acceptable limit at 400,000 unemployed, but in fact, thanks to the many assistance schemes for the unemployed, the one million mark is being passed with equanimity and the government appears to feel that the critical threshold has still not been reached.

Another solution is to rely more and more heavily on temporary workers. Let us give a few examples. Certain guarantees concerning working conditions had been obtained at the Chantiers Navals Dubigeon through the action of the very powerful unions in this company. The guarantees limiting working time in the most difficult workplaces are being sidestepped by calling upon temporary workers. The same situation as regards temporary workers can be observed in the steel industry and in the nuclear energy sector, for instance by bypassing radioactivity dosage precautions. It is possible, in fact, to check the amount of radioactivity an employee of EDF, the French energy board, working in a particular department is exposed to. But why create difficulties for oneself when it is so easy to have the most dangerous tasks done by a temporary worker, whose doses of radioactivity will never be added up as he goes from one plant to another?

Working conditions cannot be separated from living conditions, salaries

The trade union has to concern itself more and more with the workers' living conditions as a whole. In fact, depending on each particular situation, the element perceived as essential will not be the same for each worker. For a woman with young children to care for and who is faced with a lack of community facilities (transportation and cultural activities), the major worry will be to return home at a reasonable hour so as to simplify her primary problem, the care of her children. On the other hand, the man who also works on 2×8 shift in the same factory has organized his life around his 2×8 shift but is 'fed up with his bosses', so working conditions and living conditions are largely interralated.

Industrial equipment cannot be judged independently of its utilization and the organization of work around it

From the worker's point of view, a given piece of machinery or a given factory is meaningful only in relation to the conditions in which it is used. For example:

A numerically controlled grinding machine or lathe could have been conceived in the most sophisticated manner, seeking the maximum comfort for the operator, which is obviously a desirable situation. If, however, this lathe or grinding machine must be used at night because the investment must be made profitable, 'the workers will not agree to be, nightly, the slaves of this wonderful machine'. The problem takes this form, for example, in the plants of Creusot-Loire, Alsthom, and the CEM, where equipment worth tens of millions of francs has been installed and built for the French electronuclear equipment project, and where the management affirms: 'There is no question of letting such equipment stand idle at night'.

Likewise, it is undoubtedly less harmful to spend every night in an armchair, in a ventilated, air-conditioned atmosphere surrounded by green plants, than in the dust and fumes of a foundry or a workshop where benzene products are handled; nevertheless, many employees appreciate this sort of paradise very little. Thus, the engineers and technicians of Control Data have decided to take action concerning night-time working.

The organization of work means the division of labour

What would the consequences be of taking into account the needs of workers in the design of a company constructed according to the principles of Taylorism; and that in spite of the now fashionable criticisms of Taylor? At present it is the principle of an absolute distinction between the design and the execution which rules the design and construction of the factory. Now, if the semi-skilled and skilled workers had available to them the techniques of analysis which would really allow them to take a position on the design of the facility in which they worked, they would no longer be categorized as semi-skilled or skilled workers. There would no longer be the hierarchic structure which is the prop of the capitalist company. This is the reason why management speaks only in modernistic terms but carries out very limited experiments to increase the independence of employees and, at the same time, fights to eliminate union representatives from the production sectors.

The nature of the product being manufactured affects the opinions the workers have on equipment and factories

The union does not want to take the responsibility for defining the best conditions for making cars, medical equipment, AMX tanks or Concordes.

The social utility of these products is not a matter of indifference. We cannot avoid the political dimension of the worker, which makes him judge very severely the waste and the manufacturing of products which seem to him to serve no other purpose than to make money as quickly as possible.

It follows from these observations that the question we have been asked is, for a union organization, largely theoretical. Because of the organized fluidity of the job market and the socio-economic conditions which prevent the worker from giving his real opinions about his work, because of the perpetual framework of repression and the dogma of the division of labour, the industrial environment can only be defined for the 'average man'. That is to say, for nobody. It is to this environment that Mr. or Mrs. X must adapt themselves.

It is on the basis of this analysis that we have defined our policy for the improvement of working conditions. Now we will give the broad lines of this policy, together with its consequences for the design of new factories.

Our position

In the logic of the preceding analysis, it is the employees who actually work with the equipment who are best able to criticize it. This is the very basis of our organization's position. This position has already been made clear with regard to repetitive work and unskilled workers in our 1972 document under the form of homogeneous groups of production. Furthermore, it is valid for all industrial activities and all categories of employees.

For our organization, this position has the following consequences.

It is not up to the trade union in a company to decide such and such a modification in the organization of production. In this field, the trade union intervenes as a sort of relay station, a place of confrontation, a sounding-box, but not as a technical design office.

Nor is it our Federation's responsibility to promote such and such a technological process. Our role is rather to define the demands that will permit the workers to preserve their living conditions.

The following points follow on from this and clearly define our point of view.

A rejection of the principle of night shifts

This does not, of course, mean, and this is well known, that we are unaware of the necessity of performing certain tasks at night. On the other hand, we do take a stand against the extension of night work for purely economic reasons.

We can give a number of examples where union action in different firms has led to an improvement in shift organization, and consequently night work and weekend work have been reduced.

In June 1973, the workers on continuous shift work in a particular firm started a union action to obtain time off for Sunday afternoons. After 10 months, which included 38 strikes on Sundays only, they obtained seven days compensatory leave to be taken as Sunday afternoons paid time off. This was an important step in the right direction to avoid weekend work.

In January 1975, at SAFE-Hagondange (the Renault steel-making subsidiary), 1200 shift workers working a three-shift rota refused to do the shift over Saturday nights. Similar concessions had been achieved by the employees at CEGEDUR Issoire and Creusot Loire. This kind of union action is being continued, and also includes workers on a two-shift rota. The aim is to achieve a 40 hour working week over five days. The local CFDT organization, which includes the majority of unionized workers, has invited the day shift or single shift workers to actively participate in these claims.

In February 1977, 500 shift workers out of 2000 started to refuse to work a sixth night (which resulted in a night shift at the weekend once every fortnight) at J. J. Carnaud-Basse Indre. A combined action involving the two main union federations (CFDT and CGT) is still going on, with a one hour spot strike at the end of each shift. The management have made a number of concessions: one extra day's paid leave for eight Sundays worked, a half-hour compensatory reduction for the night shift, and a return to a 40 hour week in the near future. However, the workers have not accepted those changes as sufficient and continue to push for further improvements.

Rejection of paced, routine, repetitive work

For us, the promise of leisure and comfort cannot compensate for this kind of job. Paced, repetitive work is doubly unacceptable because it makes the worker a mere extension of a machine. It turns him into a robot. This kind of work, which utilizes only a part of a man's physical and intellectual capacity, is the most alienating form of industrial labour. We condemn it in the same way as we do night work.

Of course we are realists. We know that socio-economic conditions do not always permit workers to reject this type of activity. Nevertheless, our union's position remains well defined.

Two examples of modification to task organization, in line with the position taken by the union on repetitive work are interesting. The first concerns a factory of the multinational electronics firm Philips. After a very long strike action, buffer stocks were introduced along the production line; this allowed the employees to leave the line for short periods without having to be immediately replaced.

This change was accompanied by a reduction in the work rate from 1.30 to 1.17. At the same time, in a factory owned by Jaeger, a number of union negotiations were started to reduce the work rate; this opened up a debate on what was the natural work rate. The result of this action was eventually a reduction in speed of working.

The workers' capacity to intervene, along with their union organizations, in everything which affects their situation in the company must be actively encouraged

This will to intervene supposes, of course, extensive training of union militants and a considerable evolution, through collective action, in the balance of power within companies. The most important condition for this to happen is an increase in freedom of access for the union to the shop floor. Then, a few important questions arise.

Do the Comités d'Entreprises really have the means to give an informed opinion on planned investments as provided for in the texts on working conditions? These committees are formed from representatives of all groups within the firm and provide a platform for discussion on the running of the factory. Do the managements of Ford, Citroën, Peugeot or Renault, or Philips in the electronic field, consult in any way whatsoever the representatives of the Federation Générale de la Métallurgie–CFDT when a new plant is being designed? Are the opinions of local union leaders considered before deciding to set up a factory which will affect from top to bottom the living conditions in a given locality?

Conclusion

Working conditions in France remain an important source of conflict on the shop floor and play a major part in union negotiations. The Federation has recently designated working conditions a priority area for action and to this end has produced an audio-visual information package which is distributed to trade union representatives. The package shows the way working conditions are evolving in the engineering and metallurgical industries. Certain critical developments encourage us to promote improvements on such points as: increasing the autonomy of the individual worker, introducing new production processes which allow operators to take rest pauses when necessary, checking on the design of new equipment and the materials used, improving work content, limiting tasks involving piecework or pacing.

In this debate, it is important that we should not lose sight of the fact that the technology itself is a relative consideration for the workers: that is to say that working conditions are dependent on the design of the workplace and the technological environment but also include the social environment, and social relations at work remain a determining factor in the well-being of the worker in the factory.

Postscript

The preceding chapters are not likely to provide answers to everyone's specific problems but are intended to draw together the views of research workers and industrialists concerning some of the observed problems of human stresses arising from industrial work. It has been shown that these arise from many sources, affect various people in various ways, often go unrecognized but nevertheless represent serious levels of stress which give rise to ill-health and loss in production.

To have identified the existence of such stresses and how some of them arise is useful. Wider dissemination of this knowledge is necessary for its wider recognition. So too is wider discussion of the problems of working stress, for all too little is known about it and the common view is still that it is up to the individual to protect himself. Certainly, neither the engineer nor the manager can contribute to this protection if he is unaware of how the problems arise, though there is every reason to believe that the good professional in either field will seek the appropriate action if he is alerted to the need. Hence more investigation is required, stimulated by experience of men in the field and pursued in both factory and laboratory by competent researchers who can deal with both empirical and fundamental studies. Neither *ad hoc* practice nor fundamental research is sufficient on its own, for a full understanding requires interchange between the two areas of experience.

As must be clear, jobs are usually created by machine designers, production engineers, and work study personnel, often working independently. If a job exists already and a new job is to be introduced in its place, the change comes about by traditional methods, used by management from its experience, together with some bargaining between management and workers to realign the job conditions to suit their interests wherever possible. The system works but is uncertain and patently does not lead to better jobs. It cannot, because there is no stage where the policy for creating

265

a better job is introduced and pursued to its logical conclusions with the same vigour as the other factors involved in the job's make-up. This is probably because there are no formal measures for the effectiveness of such a policy, particularly measures which would permit the financial advantages to be assessed.

Classical accounting methods deal with what can be measured in money terms. Recently interest has grown in developing measures for the human resources of a company, 'human asset accounting'. There is another area which needs extending, also recent but not completely new, and that is cost-benefit assessments of ergonomics, work design, work structuring, or whatever. This may need to recognize that the improved stability of the company is a tangible asset. If the company is so delicately balanced that a short-tempered foreman, sudden change of product mix or demand for change due to unforeseen circumstances causes disruptions to production and walkouts, then this is clearly an expense. The ability to avoid this by a stable social balance in the factory is an asset which is profitable—or potentially profitable.

Such methods of analysis should provide us with other criteria for the success of our changes in production methods. They would complement the purely human measures and the performance measures to present a more complete view of the accruing benefits of activities such as those given in these pages.

What of the problems of the future, which in many cases are also those of the present? The microprocessor presents to many a threat, although it may be less than the more doom-laden prognostications would try to make us believe. In the early 1960s there were many writers predicting the automatic factory and the disappearance of conventional machines in the wake of advances in numerical control. This has not proved to be the case. What has been seen, however, is that where numerical control has not been recognized as a new manufacturing system and the jobs, skills, and organization reformed to cope with it, it has been an economic failure. The slow uptake of many new technologies in the United Kingdom may have more to do with an inability to cope with these managerial factors than any lack of perception of the technical gains to be achieved.

The so-called microprocessor revolution which is forecast will allow more information to be handled by the equipment. It is tempting to see the problems here to be those for solution by the systems analyst or programmer. But the systems analyst does not always see an operator as a human being whose role must be optimized in human terms; indeed, he will sometimes remove him from contact with the process altogether, enclosing him in a control room. Although published studies are available of good performance under such conditions there are others which demonstrate the opposite, and it is by no means certain that all operating functions can be well performed over long periods in these circumstances.

Even if computers are backed up by other computers, can the people do the jobs for which they are intended? If it is not possible to do without people reliably, then should not the jobs they will be doing be designed to suit their abilities? Then the automatic part is introduced to complete the link between the operator and the job. This puts the design engineer's task as providing a complementary link between man and process rather than the operator himself being the complementary element.

If this is to be done, the designers have to be in the original discussions regarding the roles of the operators. To provide a specification is not enough because it will be interpreted by the current philosophy, i.e. design the man out of the system as far as possible. These discussions, which should include those other ergonomic areas of the environment, the workplace layout and so on, should be right at the beginning of the factory design, because decisions here influence the plant engineering. There is usually a need for a budget for technical research; there is certainly an equal need for an ergonomics research budget. Just as a technical specification is needed, so is a social specification of the resultant plant, so that an adequate design of the work situation can be produced as reliably as the technical objectives can be achieved.

Arising from this discussion, it becomes evident that a trend towards better performance and safer and more satisfactory jobs is also a trend towards more interactive planning and management. The interactions arise where information, knowledge or interest boundaries interlink. In a manufacturing system, where people are heavily involved in all stages, the knowledge of how each stage works cannot lie entirely with management; the experience of working the process is vital knowledge. So a greater involvement and greater say by each individual in his job arise naturally from a concern jointly with efficient and humanly designed work; there is no artificial component of 'industrial relations' or 'worker democracy'. Lisl Klein (1978) expressed it by saying that 'Production engineers create industrial relations'. By appropriate design of the technology and organization to take into account the needs of those who work in industry, industrial relations can be reduced to the day-to-day relationships between those concerned with work in industry rather than remaining the major industrial headache they are today.

Reference

Klein, L. (1978). The production engineer's role in industrial relations. *The Production Engineer*, **57**, 12, 27–9.

Index

269